Acknowledgements

We thank the following people for their helpful contributions to the preparation of this report.

Professor R J Davies, St Bartholomew's Hospital, London

Dr K Venables, Brompton Hospital, London

Mr P Kendall, Department of Health

Dr M Williams, Department of Environment

Dr S Coster, Department of Environment

Permission to use the figures listed below has been applied for:—

Figures 4.1, 4.3, 4.4, 4.5, 4.6, 4.7, A4A.1, A4A.2

DEPARTMENT OF HEALTH

Committee on the Medical Effects of Air Pollutants

ASTHMA

AND OUTDOOR AIR POLLUTION

Chairman: Professor S T Holgate

Chairman of the Sub-Group on Asthma and
Outdoor Air Pollution: Professor H R Anderson

London: HMSO

ISBN 0 11 321958 X

ISBN 0 11 321958 X

Contents

Chapter 1

Executive Summary

1.1 The Committee on the Medical Effects of Air Pollutants (COMEAP) was asked by the Department of Health (DH) to advise on the possible links between outdoor air pollution and asthma, excluding biological pollutants such as pollen. A Sub-Group was set up to review this area in detail and to draft the report. The report has been agreed by the Committee.

1.2 The following terms of reference were provided by the Department.

To advise on:

(a) The time trends and geographical pattern of asthma in the United Kingdom (UK) and the relationship of air pollution to such trends and patterns.

(b) The role of air pollution in aggravating existing asthma.

(c) The possible mechanisms by which air pollution might cause or aggravate asthma.

(d) Gaps in relevant information.

(e) Recommendations for further work.

1.3 It was recognised that, while this report would focus on outdoor pollution, other forms of air pollution, including that found indoors (where most people spend the majority of their time), or associated with cigarette smoking, might also be relevant to the causation of asthma.

1.4 Asthma is a disease of the lungs in which the airways are unusually sensitive to a wide range of stimuli, including inhaled irritants and allergens. This results in obstruction to airflow which is episodic—at least in individuals with early or mild asthma—and which causes symptoms of tightness and wheeziness in the chest.

1.5 There has been an increase of about 50% in the prevalence of childhood asthma over the last 30 years, which corresponds to an increase in atopic diseases generally over this time. There has been at least a ten-fold increase in hospital admissions for asthma among children, which may partly reflect changes in medical practice.

1.6 Over the period during which asthma has been increasing, emissions of coal smoke and sulphur dioxide have fallen markedly while those of oxides of nitrogen and volatile organic compounds from motor vehicles have increased. During this time emissions of particles from coal smoke have fallen, whilst those from diesel vehicles have increased.

1.7 Data on trends and geographical variations in exposure to ozone, nitrogen dioxide and particles from vehicles are limited. The occurrence of ozone episodes in summer has probably increased over this century, but in the 15-20 years since measurements began, there is no clear trend in annual average concentrations. Annual average nitrogen dioxide concentrations have not increased in large urban centres, although there is some indication of a small increase in other urban areas.

1.8 It has been suggested that environmental factors such as air pollution could *initiate* asthma in previously healthy individuals or *provoke or aggravate* asthma symptoms in those who are already asthmatic.

1.9 While there is laboratory evidence that air pollution could potentially have a role in the initiation of asthma, there is no firm epidemiological or other evidence that this has occurred in the UK or elsewhere.

1.10   While there is some epidemiological evidence that air pollution may provoke acute asthma attacks or aggravate existing chronic asthma, the effect, if any, is generally small and the effect of air pollution appears to be relatively unimportant when compared with several other factors (eg, infections and allergens) known to provoke asthma.

1.11   There is some laboratory evidence that exposure to the common gaseous pollutants can enhance the response of asthmatic patients to allergens, though the effect does not seem to be large. There is no direct evidence for such an interaction as a result of exposure to outdoor air pollution in the UK.

1.12   There is no consistent relationship between trends in the prevalence of asthma and trends in emissions or ambient concentrations of air pollutants. A number of equally, if not more, plausible explanations for the trends in asthma have been hypothesised.

1.13   The epidemiological evidence concerning the short term effects of air pollution on asthma indicates that:

(i)   Day-to-day variations in air pollution are likely to have a small effect on the lung function of asthmatic adults and children. In general these changes are unlikely to cause symptoms. However, patients with severe asthma may be more affected because of their lower reserve of lung function. The main effects are observed in the elderly with chronic obstructive lung disease (which includes asthma).

(ii)   Seasonal patterns of asthma bear little or no relationship to those of air pollution.

(iii)   Based on studies from overseas, it is likely that the short-term fluctuations in levels of air pollution currently encountered in the UK are responsible for small changes in the numbers of hospital admissions and accident and emergency attendances for asthma. Limited experience from the UK during well defined air pollution episodes indicates that admissions may be increased by a small amount, along with similar increases in admissions for other respiratory diseases.

1.14   The epidemiological evidence concerning the geographical distribution of asthma indicates that:

(i)   There is little or no association between the regional distribution of asthma and that of air pollution.

(ii)   Prevalence studies comparing high with low pollution areas have not found consistent associations between outdoor air pollution and asthma prevalence.

(iii)   There is no convincing evidence that asthma is more common in urban areas than in rural areas of the UK. Limited evidence from the UK and other countries suggests a modest relationship between asthma prevalence and local traffic density. The extent to which this is due to air pollution has yet to be determined.

1.15   A number of recommendations for further work are made.

Conclusions

1.16   As regards the initiation of asthma, most of the available evidence does not support a causative role for outdoor air pollution. (This excludes possible effects of biological pollutants such as pollen and fungal spores.)

1.17   As regards worsening of symptoms or provocation of asthmatic attacks, most asthmatic patients should be unaffected by exposure to such levels of non-biological air pollutants as commonly occur in the UK. A small proportion of patients may experience clinically significant effects which may require an increase in medication or attention by a doctor.

1.18   Factors other than air pollution are influential with regard to the initiation and provocation of asthma and are much more important than air pollution in both respects.

1.19   Asthma has increased in the UK over the past thirty years but this is unlikely to be the result of changes in air pollution.

Chapter 2

Introduction

Asthma, a disease of multiple causes

2.1   Asthma is a matter of serious concern in Britain and most other developed countries. Currently, in Britain, about 10% of children have a diagnosis of asthma and a further 5% have asthmatic symptoms, but have not been diagnosed. About 4-6% of adults have diagnosed asthma and an unknown, but probably substantial, proportion remain undiagnosed. In its acute form, it is responsible for much distress and may threaten life. As a chronic disease it interferes with normal activities of living with consequent effects on school, work and life-style. It imposes a heavy burden on the health service at both general practice and hospital levels, and accounts for a large share of prescriptions. The total costs of asthma to the UK approach £1 billion per annum. Death rates are very low in children, but rise with age. About 2,000 deaths per year are recorded, half of which are over the age of 75.

2.2   Concern about asthma has been further increased by evidence that prevalence may be increasing.

Figure 2.1   **Time trends in indicators of asthma among British schoolchildren, 1962-92**

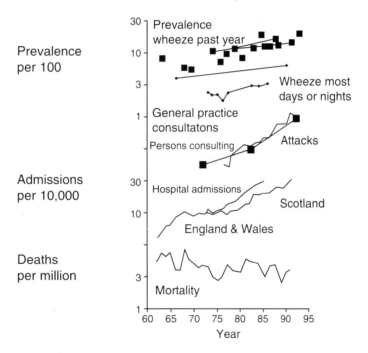

In Britain, the proportion of children reported to have had wheezing symptoms in the past year has increased by about 50% in relative terms over the past 20 or so years. The trends occurring in Britain have also been observed in many other countries, most notably in Australasia. Trend data are not available for adults in Britain, but evidence from Swedish and Finnish army recruits indicates that there has been an increase in young adults also. The use of hospital and primary medical services and the prescribing of asthma-therapies have increased, most markedly in children. In children under the age of 5, for example, hospital admission rates increased 13-fold between 1962 and 1985. Trends in the use of services have probably been influenced considerably by changes in medical practice but it is not known how much of the increase is explained by an increase in the prevalence or severity of the disease itself.

In children and young adults there has been little long-term change in mortality rates over the last 2 decades. In older persons (amongst whom mortality is highest), there has been an upward trend, but this may be an artefact due to a shift in diagnostic practice rather than a true increase.

2.3   To an asthma sufferer, the characteristic manifestation of the disease is chest discomfort and difficulty in breathing, commonly expressed as "chest tightness", which is often associated with a wheezing sound due to air being forced through narrowed airways. These symptoms may be accompanied by cough and phlegm production. Severity may range from almost imperceptible symptoms to intense distress. It is also characteristic that symptoms tend to be episodic, though some asthmatic patients may experience persistent symptoms. These symptoms result from narrowing of the airways due to swelling of the airway lining, constriction and enlargement of the surrounding muscle and secretion of fluid into the bronchial passages. It is now considered that the basic pathological abnormality is a (non-infective) inflammation of the airways. The airways of asthmatic patients tend to be more sensitive (hyperresponsive) than normal to a variety of stimuli, including allergens, chemicals, physical agents and infections.

2.4   In practice, the definition of asthma for clinical and epidemiological purposes presents problems. This is because the symptom of wheezing may also occur with other conditions, especially airways infections (such as acute bronchitis and bronchiolitis) in young children, and chronic obstructive pulmonary disease in older adults. Although, for population studies, questionnaires are more or less standardised, they tend to misclassify some other respiratory conditions as asthma, and are not good at grading severity. Criteria for a clinical diagnosis of asthma may be difficult to apply in population surveys and it has been shown that, especially in the older age groups, there is considerable diagnostic overlap with chronic obstructive pulmonary disease at all levels of care, and in death certification.

2.5   The measurement of asthma in population studies employs the related concepts of incidence, prognosis and prevalence. Incidence is the rate of occurrence of asthma over a given period and has two separate aspects. Incidence ("persons") is the rate at which new cases of asthma occur in the population, ie, the rate at which previously normal individuals become asthmatic. Factors responsible for incidence ("persons") shall be referred to in this report as initiators. Incidence ("spells") is the occurrence of asthma attacks in persons who already have asthma. Factors responsible for incidence ("spells") are called provoking, aggravating, precipitating or triggering factors. In this report we shall mainly use the term "provoking" factors. The distinction between these two forms of incidence is of fundamental importance in considering the effects of air pollution on asthma. Prognosis is the probability that an individual with asthma will still have the disease at a given time in the future. It is a measure of the persistence or otherwise of the disease. Prevalence is the proportion of the population with asthma at a particular point or over a given period of time (eg, in the 12 months before interview). It is the easiest parameter to measure using cross-sectional survey techniques and consequently, one for which there are most data. However, it must be recognised that prevalence is a function of both types of incidence and also of prognosis. The severity of asthma is difficult to define in a standardised way. It is a function not only of the intensity of attacks, but also of their duration and frequency. Methods of measuring severity include lung function tests and questionnaires.

2.6   The most important known risk factors for the *initiation* of asthma are a family history of asthma, evidence of atopy, or the existence of other atopic disorders—allergic rhinitis or eczema. In this context, atopy is defined as a tendency to develop allergic antibodies to environmental agents; this can be assessed by blood tests or allergen skin tests. Otherwise, little is known about the factors which are responsible for induction, though there are a number of theories involving diet, infection and pollutants, including cigarette smoking, possibly operating early in life. There is widespread concern that air pollution may play a role in the induction of asthma.

There is evidence that other atopic diseases such as allergic rhinitis and eczema are also increasing; this points towards the possibility that the increase in asthma is related, in whole or in part, to an increase in the prevalence of atopy.

2.7   Much more is known about the provoking agents of asthma. These include: inhalation of smoke, fumes, chemicals and allergens; respiratory infections, especially of a viral type; certain foods; exercise; and emotional stress or excitement. It is widely believed by the general public and doctors that air pollution, especially that from traffic, is responsible for the inciting of attacks or worsening of existing asthma.

2.8   There is no evidence that any particular environmental agent is a "necessary" cause of the initiation or provocation of asthma (ie, no evidence that asthma cannot occur without a particular factor). This means that the patterns of causes of asthma and of asthma attacks are likely to vary between persons, populations, places and over time. Many of the known and potential causal factors in the environment or associated with lifestyle may be changing with time. Air pollution is but one of these factors and, for this reason, the role of air pollution needs to be considered in relation to other competing causes of asthma and their changes over time.

Air pollution, a possible cause of asthma

2.9   Air pollution refers to the presence in the air of gases or particles of matter, which are not natural to the atmosphere at such concentrations. These are generally man-made, though the term also applies to unusual natural events, such as the eruption of volcanoes. This report will be confined to the effects of non-biological pollutants. It will exclude aeroallergens (pollens, mould spores) but it is recognised that aeroallergens are important causes of respiratory problems (including outbreaks of asthma) and, being to some extent influenced by man, could be included as pollutants. The possibility of interactions between aeroallergens and air pollution is important and will be discussed in this report.

2.10   In seeking an explanation for the apparent increase in asthma it is entirely reasonable to examine the hypothesis that air pollution may play a part, since adults breathe about 20,000 litres of air per day, and the lungs are consequently in intimate contact with the surrounding atmosphere.

2.11   Smog episodes occurring as a result of pollution from the burning of coal in stagnant winter weather conditions have long been thought to be a cause of illness and death. This was proved conclusively during the 1952 London smog when there was a great excess of deaths, especially from respiratory conditions in older persons. At the other end of the scale, it is a common experience that exposure to fumes or smoke, for example from bonfires, may irritate the lungs even of healthy persons. It would therefore seem likely that a condition such as asthma, which is characterised by hyperreactivity of the airways, would be affected by air pollution. This is the experience reported by many asthmatic patients.

2.12   The main ambient pollutants which have been studied in relation to respiratory diseases are particles, sulphur dioxide (SO_2), nitrogen dioxide (NO_2) and ozone (O_3). Full accounts of the properties of these pollutants and their trends and distribution in Britain may be found in a number of recent reviews and these aspects will not be covered in detail here. However, the main trends and patterns of air pollution will be summarised and further data will be presented in the chapters concerned with the public health effects of air pollution.

2.13   Emissions of SO_2 have declined since the introduction of the Clean Air Acts of 1956 and 1968 and as a result of a change to less sulphurous coal and the widespread introduction of natural gas (see Figure 2.2). Furthermore, emissions from power stations using fossil fuels now tend to be away from concentrations of population. High stack emissions with efficient dispersion has, however, led to a wide exposure to low levels of pollutants. Belfast, where coal and smokeless fuel are still burnt in domestic grates, is the remaining major city to experience poor SO_2 conditions. High levels also occur in some parts of northern England where domestic consumption of coal remains high. Contributions by various sources to total UK SO_2 emissions are shown in Figure 2.3.

Figure 2.2 **UK emissions of sulphur dioxide 1970-1993**

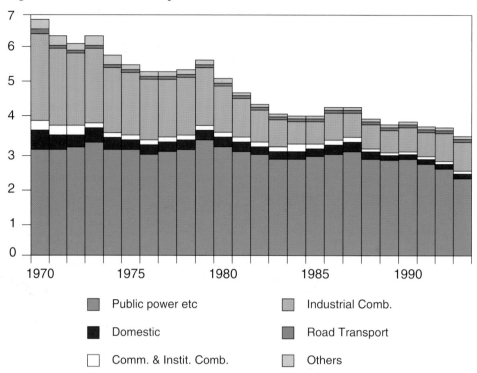

Figure 2.3 **UK emissions of sulphur dioxide 1993. Total emission = 3.2 MTonnes**

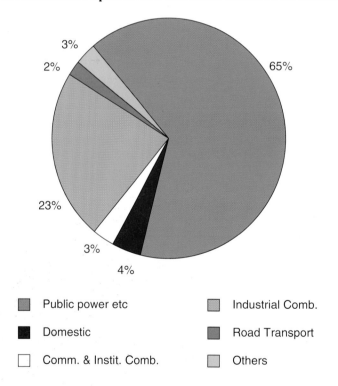

2.14   Emissions of Black Smoke have fallen markedly since the 1950s (see Figure 2.4). This is due to the reduction of coal burning in domestic grates and the increased combustion efficiency of power and industrial sources. The downward trend in emissions of particles has, however, now ceased. This is because of an increased contribution from road traffic, mainly diesel engines which emit small carbon-rich particles. The main measure of particles in use in the UK has been a smoke stain method known as Black Smoke. Levels of Black Smoke in cities such as London have fallen markedly (see Figure 2.5) and, like SO$_2$, are now at an historically low level. The Black Smoke technique is dependent on the colour of the particles and this has changed with the changing emission sources. It is important to note that chemical particles such as sulphuric acid aerosols and sulphates, which are also present, are not measured by this method. A newer gravimetric method of measuring particles (PM_{10}) has not been used long enough in the UK to provide a clear indication of trends. Particles are a complex mixture of different substances of varying sizes and this may have implications for potential health effects.

Figure 2.4 **UK emissions of Black Smoke 1970-1993**

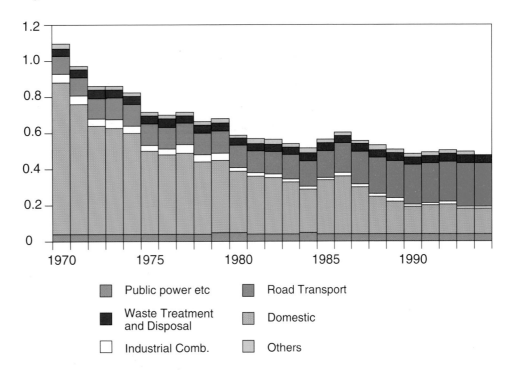

- ☐ Public power etc
- ■ Waste Treatment and Disposal
- ☐ Industrial Comb.
- ☐ Road Transport
- ☐ Domestic
- ☐ Others

Figure 2.5 **98th percentile of daily concentrations of Black Smoke at "Stepney 5"**

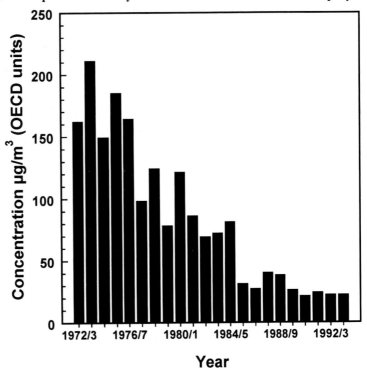

2.15   Oxides of nitrogen (NO_x) are produced in high temperature combustion conditions such as are found in petrol and diesel engines. Emissions have increased, with an increasing proportion coming from mobile sources. Trends in emissions are projected to stabilise as a result of the introduction of three-way catalytic converters, but to rise again as traffic volume increases (see Figure 2.6). At the tail pipe, only about 10% of NO_x is NO_2, the remainder being mainly nitric oxide (NO), which is not thought to be a danger to health. In the ambient air, NO is oxidised to NO_2 which is regarded as potentially harmfully to health. Exposure to NO_2 is related to traffic density. In Central London, there is evidence of an upward trend in the peak concentrations of NO_2 (measured as the 98th percentile of hourly values) over the past 20 years (see Figure 2.7). Other sites show a stable concentration with time, but this may be explained by traffic saturation. In December 1991, during a period of stagnant winter weather, London experienced historically high levels of NO_2 (423 ppb, 795 $\mu g/m^3$) which exceeded WHO guidelines by a factor of 2. Other cities, such as Manchester, have also experienced similar episodes (369 ppb, 694 $\mu g/m^3$, NO_2 in December 1992).

Figure 2.6 **UK road transport emissions of NO_x**

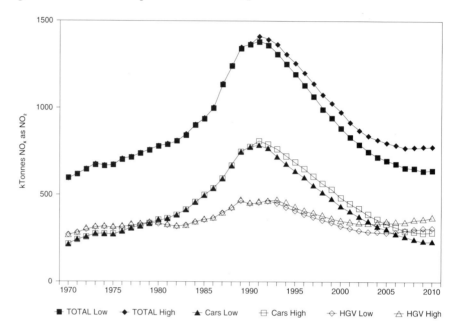

2.16   Ground level (tropospheric) ozone is formed mainly as a "secondary" pollutant by the action of sunlight on NO_2 and oxygen in the presence of volatile organic compounds and hydrocarbons which act as catalysts and are also emitted by combustion sources such as motor vehicles and industrial processes.

$$NO_2 + h\upsilon \rightarrow NO + O^\bullet$$
$$O_2 + O^\bullet \rightarrow O_3$$

Concentrations of O_3 tend to be lower near to sources of emission of oxides of nitrogen because NO reacts with O_3 to form NO_2. In certain weather conditions in the summer, many people may be exposed to increased levels of ozone. The occurrence of ozone episodes in summer has probably increased over this century, but in the 15-20 years since measurements began, there is no clear trend in annual average concentrations. There is no evidence for an increase in peak (hourly) ozone concentrations over the past 15-20 years (see Figure 2.8). Sporadic episodes, such as those occurring in 1976 and to a lesser extent in 1990 in southern England, tend to reflect the variable nature of English summer weather.

2.17   There are many other pollutants in the ambient atmosphere (eg, aldehydes, methane, ammonia, carbon monoxide and organic compounds such as benzene) but, because little is known about their distribution or possible relationships with asthma, these will not be considered in this report. A wide range of pollutants of biological origin are also found in the air. These include pollens, mould spores, mites and dander. The importance of these pollutants in asthma is well recognised and will not be dealt with in detail in this report. Here the emphasis is on outdoor, chemical air pollution. However, it is important to note that because the population is exposed to mixtures of pollutants, not to pure substances, any associations observed with asthma may not be due to the pollutants measured most commonly (indicator pollutants) but to another substance or substances with which they are associated, or to an interaction between two or more pollutants.

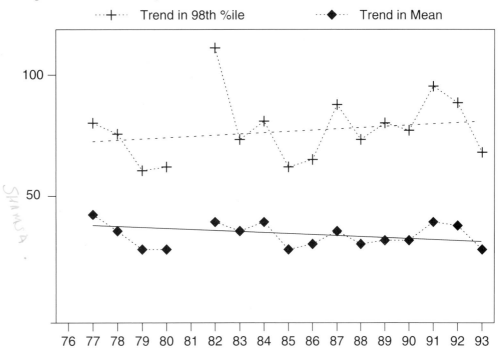

Figure 2.7 **Trends in NO₂ emissions, 1976-1993**

2.18   More time (> 80%) is spent indoors rather than outdoors. To an extent which varies from pollutant to pollutant, outdoor levels influence indoor levels, and outdoor climate influences indoor climate. Some outdoor pollutants, such as fine particles and NO_2, easily penetrate indoors. Others, notably O_3, are so reactive with surfaces such as those of furnishings, that levels are low indoors. The indoor environment itself is a source of various pollutants, some being the same as found outdoors. Exposure to NO_2 occurs in the home due to the use of gas cookers and may reach levels which exceed those in the outside environment. Indoor exposures to other agents such as allergens, organic dusts and environmental tobacco smoke should be considered as competing causes of asthma, though these are not dealt with in this report.

Figure 2.8   **Trends in peak ozone concentrations, 1977-1993**

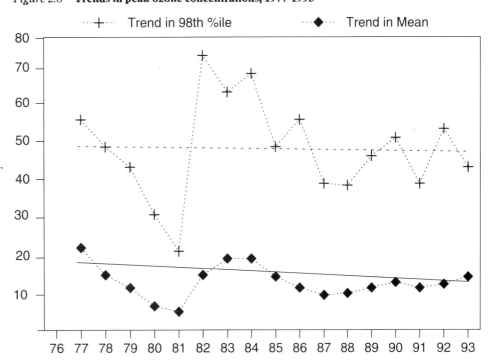

Asthma and air pollution—are they related?

2.19 The apparent correspondence between trends in asthma and trends in some types of air pollution, from traffic in particular, is one basis for the present concern. This has been strengthened by the emergence of laboratory studies which have pointed to the theoretical possibility that air pollution could cause respiratory problems, including the provocation of asthma and nasal allergies. This hypothesis has been supported by a number of studies of ambient air pollution among exposed populations. The current evidence applies mainly to the role of air pollution as a provoker of these diseases rather than to their induction among previously healthy subjects.

2.20 At the same time, there has been increasing concern about environmental issues generally, and about the growth of traffic in particular. There is recognition that present trends in energy use are not compatible with preservation of the natural environment either locally, or globally. Environmental issues have become politically important and various pressure groups have been very influential in raising public concern about the environment. In doing so, the possible connection between air pollution and health, especially asthma, has become a frequent news item, as well as being the subject of a number of television and newspaper documentaries.

2.21 This public concern has been shared by government, and through the Departments of Health and Environment (DH and DoE), a number of committees and working parties have been set up to examine the evidence concerning air pollution and health with a view to making recommendations concerning air quality standards, issuing medical advice during air pollution episodes and suggesting research priorities. The Committee on the Medical Effects of Air Pollutants (COMEAP) was constituted in 1992 with the following remit:

(a) To assess, and advise Government on, the effects upon health of air pollutants both in outdoor and indoor air, and to assess the adequacy of the available data and the need for further research.

(b) To co-ordinate with other bodies concerned with the assessment of the effects of exposure to air pollutants and the associated risks to health and to advise on scientific discoveries relevant to the effects of air pollutants upon health.

An outcome of one of the early meetings was to set up an Asthma Sub-Group comprising some members of COMEAP and co-opted experts, to advise the Committee on the matter of air pollution and asthma. The terms of reference of the Sub-Group on Asthma and Outdoor Air Pollution were to advise on:

(a) The time trends and geographical pattern of asthma in the UK and the relationship of air pollution to such trends and patterns.

(b) The role of air pollution in aggravating existing asthma.

(c) The possible mechanisms by which air pollution might cause or aggravate asthma.

(d) Gaps in relevant information.

(e) Recommendations for further work.

Method of working and structure of the report

2.22 Members of the Sub-Group were aware of a number of existing reviews of the evidence concerning asthma and air pollution, but had noted that none of these attempted to examine the evidence comprehensively. In particular, the citation of public health evidence was often unsystematic. To avoid the potential problem of selective citation and consequent bias, the Sub-Group decided that an attempt to examine all the available evidence should be made.

2.23 The structure of this report will reflect the range of scientific evidence available up until May 1995, from laboratory experiments to studies among whole populations. Where it is possible to conduct controlled experiments, as is usually the case in laboratory studies, cause and effect relationships are usually clear. The problem then is in extrapolating from these findings to effects in exposed populations. In contrast, when examining the relationship between air pollution and asthma in mobile populations using observational rather than experimental methods, cause and effect may be difficult to determine because other factors may be confounding the association and, even if these are known, they may be impossible to allow for. Thus, as studies become more relevant to public health, they become more difficult to interpret.

2.24   In Chapter 3 an overview of the clinical and immunological features of asthma is provided. It was felt that such an account was needed to set the evidence presented in Chapters 4-9 in context.

2.25   Chapter 4 will address the following question:

"Are there plausible mechanisms by which air pollution could induce or incite asthma?"

It will review the relevant physiological, biochemical and immunological information concerning the ways in which the lung defences handle gaseous and particulate substances, and the potential for these pollutants to interact with basic mechanisms underlying asthma. This will include the question of interactions between air pollution and aeroallergens. Much of the evidence will be based on animal experiments, with the consequence that its relevance to the human situation may be unclear.

2.26   Chapter 5 will describe the evidence from controlled experiments using adult volunteers. These are often described as "chamber studies" referring to the usual method of exposing subjects. It will address the following questions:

1)   *Can short term exposure to inhaled pollutants under controlled experimental conditions make asthma worse?*

2)   *Can prior exposure to air pollutants increase the response of the lung to inhaled allergens?*

Chamber studies have usually examined the short-term individual or combined effects of gaseous pollutants (NO_2, SO_2 and O_3) on adults. For practical reasons, including the great difficulty in reproducing in the laboratory anything close to the ambient particulate mixture, very few studies, excluding those involving exposure to acid aerosols, have been conducted using particles. Ethical considerations preclude studies in children. Measures of effect include symptoms, respiratory function tests, bronchial hyperreactivity or indicators of inflammatory changes in the airways. These studies are comprehensively reviewed in four reports on air pollution by the Department of Health and will only be summarised in this report. The results of such studies carried out using asthmatic patients are directly relevant to the question of whether asthma is made worse by air pollution. This technique has recently been used to test the hypothesis that, among asthmatic patients, prior exposure to pollutants increases the response to inhaled allergens.

2.27   The evidence from animal and human experimental studies is important for deciding whether it is plausible that air pollution could provoke or exacerbate asthma. Chapters 6, 7, 8 and 9 will address the question which is more relevant to public health:

"Does ambient air pollution have a role in the provocation or induction of asthma in the population at large, and if so, what is the scale of the effect?"

The population or epidemiological studies on which this evidence is based fall into various types according to the method used. Because ambient air pollution is a mass exposure, most epidemiological strategies examine its effects on whole populations rather than at the individual level. This contrasts, for example, with the investigation of the effects of smoking using case-control and cohort techniques within populations. Comparisons between populations may be categorised as either "temporal" or "spatial" (geographical) in concept. In temporal studies, such as panel studies or population time-series studies, the person/population is used as its own control; this minimises the confounding effects of relatively stable factors such as, for example, smoking habits, use of gas for cooking, or social class. Temporal studies provide information about the acute effects of pollution on asthma and are relevant to the question of provocation. In spatial or geographical studies comparisons are made between areas which differ in levels of air pollution or in health outcome; here, stable confounders, such as smoking etc, are a serious problem unless information about them can be collected as part of the investigation. Associations observed in spatial studies provide information about the initiation and provocation of asthma.

2.28   Chapter 6 will address the question:

"Does air pollution cause short-term effects on symptoms or lung function in panels of healthy individuals and asthmatic patients?"

In these so-called "panel" studies, groups of healthy or asthmatic individuals are closely monitored, together with their exposure to air pollution. The association between health outcomes, such as lung function, symptoms or medication use, and air pollution is usually examined at a daily level using appropriate statistical techniques. Care needs to be taken in order to control for factors which may correlate with both air pollution and the outcome, such as weather conditions. The applicability of any findings to asthmatic patients in general will depend, amongst other things, on the representativeness of the sample.

2.29   Chapter 7 will address the following questions:

1)   What are the long term time trends in asthma and other manifestations of allergy in the UK and elsewhere?

2)   Do changes in prevalence over time correspond to changes in air pollution?

One of the main causes of public concern is a possible link between an increase in asthma and changes in air pollution. Correlations over time can indicate whether an association is consistent with the hypothesis but cannot provide firm evidence of cause and effect. Most of the data to be reviewed will have been obtained with other purposes in mind, but available evidence from planned studies in which changes in prevalence have been correlated with improvements or deteriorations in air quality will also be reviewed.

2.30   Chapter 8 will examine the relationship between short- and medium-term variations in asthma with those of air pollution and asks the following questions:

1)   Do seasonal variations in asthma correspond to seasonal variations in pollution?

2)   Are daily variations in asthma morbidity associated with daily variations in levels of air pollution?

Seasonal factors are important determinants of the frequency and severity of asthma. There are several possible explanations for seasonal effects, including respiratory infections, indoor and outdoor allergen exposure and exposure to air pollution. Day-to-day variation is also a feature of asthma and the contribution of air pollution to this variation has been a subject of considerable research. Where associations are observed, it may be difficult to infer causality because of the close relationships between air pollution and other weather-dependent variables such as pollen levels and temperature.

2.31   Chapter 8 continues with temporal relationships by examining the evidence from unusual events:

1)   Are asthma epidemic days caused by air pollution?

2)   Are air pollution episodes associated with an increase in asthma?

In contrast to the daily time-series studies reviewed in Chapter 8, where the effects of air pollution, if any, are likely to be small, these two questions relate to major environmental and health related events. Associations observed during such events will be important for assessing the potential of air pollution to cause acute asthma.

2.32 Chapter 9 will be concerned with spatial associations between air pollution and asthma and other allergic diseases. It will address 5 questions:

1) *Do regional variations in asthma correspond to variations in air pollution within the UK?*

2) *Is the prevalence of asthma associated with air pollution in other countries?*

3) *Is the prevalence of asthma higher in urban than in rural areas?*

4) *Is the prevalence of asthma related to exposure to traffic?*

5) *Are variations in prevalence within the united Germany consistent with an effect of the "new" pollution on asthma?*

Sections 1 and 2 will address regional and international variations which might reflect differences in the induction or provocation of asthma; it will not normally be possible to distinguish these. Section 3 will address the belief and concern that asthma is more common in urban areas, possibly because of air pollution. Section 4 will address this question more specifically by reviewing all the studies of asthma and proximity to traffic. Lastly, in Section 5, comparisons of prevalence between the former East and West Germany will be examined. This is very relevant because the former East Germany was, until recently, exposed to smoke and SO_2 pollution similar to that which was prevalent in Britain in the two post-war decades. By contrast, the former West Germany has been exposed for a longer period of time to traffic related pollution as now experienced in Britain.

2.33 The membership of the Sub-Group, as listed in Appendix 4, is intended to provide a balance between clinical, epidemiological and basic scientists involved in work on asthma. In the course of preparing this report, informal discussions have also been held with other experts in relevant fields.

Bibliography

Burney PGJ. Epidemiology. In: Clark TJH, Godfrey S, Lee TH, editors. Asthma. London: Chapman & Hall Medical, 1992; 254-308.

Commission of the European Communities. Air Pollution Epidemiology Reports Series. Report No 4. Study Designs. COST 613/2. (EUR 15095 EN). Luxembourg: Office for Official Publications of the European Communities, 1993.

Department of Health. Asthma: an epidemiological overview. Central Health Monitoring Unit Epidemiological Overview Series. London: HMSO, 1995.

Department of Health. Advisory Group on the Medical Aspects of Air Pollution Episodes. First Report: Ozone. London: HMSO, 1991.

Department of Health. Advisory Group on the Medical Aspects of Air Pollution Episodes. Second Report: Sulphur Dioxide, Acid Aerosols and Particulates. London: HMSO, 1992.

Department of Health. Advisory Group on the Medical Aspects of Air Pollution Episodes. Third Report: Oxides of Nitrogen. London: HMSO, 1993.

Department of Health. Committee on the Medical Effects of Air Pollutants. Non-Biological Particles and Health. London: HMSO, In press.

Fletcher RH, Fletcher SW, Wagner EH. Clinical epidemiology: the essentials. 2nd Edition. Baltimore: Williams and Watkins, 1988.

Lenney W, Wells NEJ, O'Neill BA. The burden of paediatric asthma. Eur Respir Rev 1994; 4:49-62.

Lung and Asthma Information Agency. Factsheets on asthma and respiratory disease. London: St George's Hospital Medical School.

Quality of Urban Air Review Group. Urban Air Quality in the United Kingdom. First Report. London: QUARG, 1993.

Quality of Urban Air Review Group. Diesel vehicle emissions and urban air quality. Second Report. London: QUARG, 1993.

Strachan DP. Epidemiology. In: Silverman M, editor. Childhood asthma and other wheezing disorders. London: Chapman & Hall Medical, 1995.

United Kingdom Photochemical Oxidants Review Group. Ozone in the United Kingdom 1993. London: Department of the Environment, 1993.

Chapter 3

A General Review of Asthma

Introduction

3.1   There is a common perception in many countries that there has been an increase in the prevalence and severity of allergic diseases and particularly asthma. Epidemiological evidence to support this has been slow to accumulate and contentious, mainly because there has been a lack of an agreed working definition of the condition. Without a satisfactory definition, it becomes impossible to assess whether there have been real changes in the prevalence of the disease and to associate such changes with any variations in the environment.

3.2   Concepts about what constitutes asthma range from descriptions of symptoms through to precise histopathological diagnosis. It is becoming very clear in children that there are at least two phenotypes associated with recurrent coughing and wheezing which might be termed asthma and the same may well be true in later adult life. The most precise definitions apply only to atopic disease associated with recurrent coughing and wheezing.

The hospital physician's view of asthma

3.3   The International Consensus Report of the American National Heart, Lung and Blood Institute has produced a definition of asthma which includes both histopathological and clinical criteria. It is as follows:

> "Asthma is a chronic inflammatory disorder of the airways in which many cells play a role, including mast cells and eosinophils. In susceptible individuals, this inflammation causes symptoms, which are usually associated with widespread but variable airflow obstruction that is often reversible either spontaneously or with treatment and causes an associated increase in airway responsiveness to a variety of stimulae."

This definition is based, in part, on the histological appearance of bronchial biopsies predominantly obtained from studies in young adults with mild to moderate asthma. It is likely to be applicable to any allergic individual aged between 5 and 45 years who has positive skin prick tests and raised IgE antibodies, complains of recurrent coughing and wheezing and who has variable airflow obstruction but does not always have evidence of bronchial hyperresponsiveness. It may be less applicable outside these years.

3.4   Within this precisely defined group of patients, there is evidence that prevalence has increased, though perhaps by less than has been suggested by some epidemiological studies. The increase in prevalence of childhood asthma has been paralleled by an apparent increase in other atopic diseases, such as allergic rhinitis and eczema. As the immunopathology of atopic asthma is unravelled, it is becoming possible to identify environmental factors which contribute to the development of the disease (inducers) and factors which exacerbate pre-existing disease (provoking factors).

The respiratory physician's view

3.5   A less precise definition, not founded on histopathology, has been used over the last 30 years:

> "A condition characterised by wide variations over short periods of time in resistance to airflow in intrapulmonary airways."

This definition incorporates the idea of bronchial hyperresponsiveness and bronchospasm as the "*sine qua non*" of the condition. In the past, this was considered to be synonymous with recurrent symptoms. However, recent studies have shown that this is not the case. There are individuals in the population who have evidence of bronchial hyperresponsiveness on challenge but no symptoms suggestive of asthma. On the other hand, there are asthmatic patients who do not have evidence of bronchial hyperresponsiveness. Furthermore, within groups of patients with asthma, the degree of hyperresponsiveness can vary independently of symptoms and of changes in the results of lung function tests used to record variations in airway calibre. Other conditions which are associated with intrapulmonary airway narrowing, with or without airway inflammation, are associated with heightened bronchial hyperresponsiveness and variations in airway calibre over time. This is particularly relevant to respiratory diseases in infancy and chronic bronchitis and emphysema in middle and old age.

The paediatrician's view of asthma

3.6   The International Paediatric Asthma Consensus Group had great difficulty in agreeing an all-embracing definition of the condition to cover all children's age groups and, in particular, infancy. Therefore, pragmatism dictated the following suggestion:

> "Asthma is a condition in which episodic wheeze and/cough occur in a clinical setting where asthma is likely and other rarer conditions have been excluded."

This concept is based on the knowledge that the younger the child presenting with recurrent coughing and wheezing, the higher the probability of an alternative diagnosis which must be excluded, such as recurrent milk aspiration, cystic fibrosis, immune deficiency, primary ciliary dyskinesia syndrome, or congenital malformation. Furthermore, there is no histological information available to establish whether those who wheeze in infancy and who go on to have atopic asthma, have eosinophilic airway inflammation at the onset of symptoms. In addition, it is appreciated that only 20% of infants who wheeze and who are not found to be suffering from some other illness, go on to have established atopic asthma. Recent studies suggest that there are at least two phenotypes of recurrent coughing and wheezing in infancy: one associated with atopic asthma, which is rare below the age of 1 and increases in prevalence between 1 and 5 years of age, whilst the second is characterised by wheezing associated only with viral infections and which is most common below the age of 1 year and declines in prevalence progressively thereafter. It is possible that this second phenotype is associated with a recurrence of symptoms in late adult life.

The epidemiologist's definition of asthma

3.7   Some of the best epidemiological studies have used symptoms as the basis for identifying asthmatic individuals. In these studies reports of recurrent wheezing, defined as "a high pitched musical sound coming from deep down in the chest when breathing out", during the last year has been used to define annual period prevalence. Reports of wheezing occurring at any time have been used to define cumulative prevalence. The prevalence of wheeze in the population is very different from that of doctor-diagnosed asthma. The latter is very prone to changing fashions in diagnosis and it is clear that this has increased dramatically over the last two decades. Some epidemiological studies have also included identification of individuals with recurrent cough. It is more difficult to relate recurrent cough than recurrent wheeze to asthma though it may well be important in some individuals. Many young children have paroxysmal cough at night and on exertion as their only obvious symptom. Whether such patients form an additional phenotype independent of wheezing still requires further investigation. Wheezing illnesses in infancy affect up to 40% of the total population with a rapid decline over the first few years. Doctor-diagnosed asthma in the first few years of life amounts to 5–6% below 4 years of age in England and Wales, rising to 10% between 5 and 15 years of age.

Ontogeny of atopic asthma

3.8   Whilst hereditability forms the basis of the definition of atopy, it is clear from studies of identical twins that the environment has a considerable influence on phenotype. Nevertheless, inherent factors play an important part in influencing immune responsiveness and dictating whether an individual can be easily sensitised to allergens (see Figure 3.1). Recent research has suggested that there may be a single gene located on the long arm of chromosome 11 which determines predisposition to allergy and asthma. However, many other studies have failed to replicate this work. It is likely that there are a number of genes involved and several pre-existing

Figure 3.1 **Relationship between genetic and environmental factors in asthma**

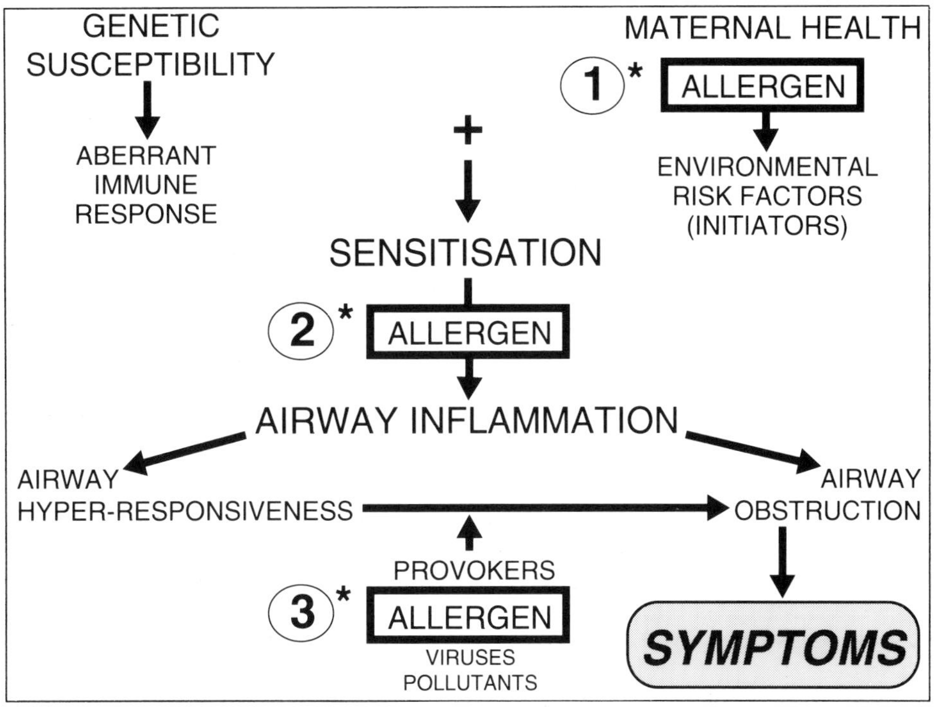

```
GENETIC                                    MATERNAL HEALTH
SUSCEPTIBILITY
                                    ①*  [ALLERGEN]
        ↓
  ABERRANT              +                  ENVIRONMENTAL
  IMMUNE                ↓                   RISK FACTORS
  RESPONSE                                 (INITIATORS)
                  SENSITISATION

              ②* [ALLERGEN]
                        ↓
              AIRWAY INFLAMMATION
        ↙                          ↘
AIRWAY                                        AIRWAY
HYPER-RESPONSIVENESS  ⟶              OBSTRUCTION
                   ↑
              PROVOKERS                     ↓
        ③* [ALLERGEN]              ( SYMPTOMS )
              VIRUSES
              POLLUTANTS
```

*points of action of allergens

immunological defects have been described in association with the subsequent development of atopy and asthma. The range of defects includes transient or permanent IgA deficiency, deficiency of the second component of complement, opsonisation defects and abnormalities of helper and suppressor T cell numbers. The role of T lymphocytes and cytokines in the regulation of IgE-mediated hypersensitivity responses to inhaled proteins has become a focus for recent research. Inhaled allergens are processed by antigen-presenting cells in the airways. These cells then present allergenic peptides to T helper cells in the airway wall or local lymphoid tissue in association with major-histocompatibility-complex molecules. The allergen specific T cells in the airways of allergic asthmatic individuals are triggered to generate a range of cytokines, many of which are pro-inflammatory. Based initially on studies in mice but, more recently, in humans, it has been shown that the T cells of atopic asthmatic patients selectively produce interleukins IL-4, 5, 6, 10 and 13 when stimulated (Th2 cells), whilst non-atopic T cell clones produce IL-2, 12, IFN-γ and tumour necrosis factor beta (TNF-β) (Th1 cells). Both clones of cells produce IL-3 and GM-CSF. It is becoming apparent that the Th2 phenotype develops very early and may antedate the manifestations of atopy, in terms of skin tests and positive IgE antibodies, and disease. The phenotype may even be present at birth and may be under the influence of factors operative during pregnancy. Those suggested include maternal smoking, maternal allergy and high allergen exposure during pregnancy, and maternal nutrition. However, all of these associations require further elaboration in prospective studies. There is absolutely no information on whether outdoor pollution may contribute to this intrauterine induction of sensitisation.

3.9 Following sensitisation, which may occur antenatally, there is a latent period before allergic inflammation becomes localised in the target organ which, for asthma, is the airways. Animal studies have suggested that there is a period of anergy to aerolised allergen in the immediate postnatal period, followed by the development of allergic sensitisation which does not occur if adult animals are exposed to the same aerolised allergen protocol. This early development of a local airway immune response following exposure to allergen has also been suggested by immuno-histochemical studies of tracheal wall from post mortem specimens in early infancy. Such observations are in keeping with those that have been made regarding month of birth in relation to allergen exposure. Thus, in Scandinavian countries, allergy to birch pollen occurs most frequently in individuals born just before the birch pollen season. It has been suggested that if month of birth could be manipulated, the frequency of birch pollen allergy could be reduced by 28% in Sweden. Similar observations in relation to seasonal variations in allergen concentrations have been made for house dust mite, grass pollen and moulds. Associations between possession of either a cat or dog in infancy and subsequent prevalence of cat or dog allergy have emphasised the importance of early allergen experience in relation to the development of allergic disease (including asthma) as distinct from just allergy. Continuing exposure to allergens and, in particular, house dust mite in temperate climates such as that of the United Kingdom, has been associated with an earlier onset of asthma and a greater probability of its continuation through into adolescence. It is likely that early allergen exposure of the target organ draws the activated lymphocytes into the airways where local inflammation can then be established if there is continuing allergen exposure. This, in turn, will produce airway narrowing and airway hyperresponsiveness.

3.10 There is virtually no information to establish whether this early process of localisation of inflammation to the airways is promoted by exposure to pollutants or, indeed, any other factors. There is, in fact, some suggestion that virus infections do not play a part at this stage in the ontogeny of the response. Once airway inflammation has been established, there is a host of factors which can provoke symptoms: and these include further allergen exposure, virus infections and air pollutants.

3.11 The likelihood that an allergic state will be produced is dependent upon three factors:

(a) degree of atopic susceptibility;

(b) severity of exposure to allergen (dependent on level of allergen in air and the duration of exposure);

(c) potency of the allergen.

The exact quantitative relationship between these factors and the likelihood of an allergic state being induced is not well understood. The degree of atopic susceptibility depends on genetic factors, possibly age, and perhaps upon the presence of adjuvant factors. The latter may include exposure to environmental tobacco smoke, particularly antenatally. Whether pollutants or infection are involved, remains to be established. The importance of the level of allergen to which the individual is exposed is well established for occupational allergy, and has also in infancy, in relation to seasonal allergens and pets, been demonstrated. The concentration of exposure at an airway level is appreciably influenced by particle size. The potency of the allergen relates not only to molecular structure, but also to the rapidity with which proteins are released by inhaled particles. Thus, house dust mite faecal particles and pollen grains release enzymes very rapidly when they alight on a moist mucosal surface. This is not the case for mould spores which are relatively inert and are consequently associated with a very much lower prevalence of allergy, despite relatively high exposure of all individuals, both in the domestic and outdoor environment. The ease with which allergens are released by particles may be increased by pollutants. It has, for instance, been suggested that association of pollen grains with diesel particles might result in disruption of the grain and release of allergenic proteins which would then more easily sensitise the airway. The only epidemiological evidence to support this contention comes from Japan in relation to cedar allergy (see Chapter 9).

Wheezing, asthma and chronic obstructive lung disease

3.12 The relationship between wheezing in infancy and asthma and chronic obstructive lung disease in late adult life remains obscure. Whilst prospective studies have begun to disentangle the relationships between infancy wheezing and childhood atopic asthma, only retrospective studies will be able to define a relationship with later onset respiratory disease. It is clear that atopy is the most important factor predisposing to the persistence and increased severity of wheezing in childhood. However, wheezing in infancy is a poor predictor of atopic asthma at 5–10 years of age. This suggests a different mechanism for airflow limitation in infancy with symptoms more related to airway size than airway inflammation. Maternal smoking during pregnancy and low birthweight are major factors increasing the frequency of wheezing illness in infancy, and more recently, these symptoms have been shown to be associated with smaller airway size. As obstructive disease in late adult life has been associated with impaired fetal and infant growth, there may well be a link between infant wheezing and chronic obstructive lung disease in late adult life. There is also a possible link between abnormalities of fetal growth and a greater propensity to developing IgE antibodies. Thus, there may be factors common to both the development of atopy and unusually narrow airways, though the pattern of fetal growth associated with atopy is somewhat different to that associated with later-onset lung disease.

3.13   It is clear that maternal cigarette smoking in pregnancy affects fetal growth and is one of the major factors that has been related to wheezing illnesses in infancy. Whilst exposure to environmental tobacco smoke in pregnancy has also been associated with higher IgE levels and predisposition to atopy, the majority of infant wheezing associated with maternal cigarette smoke exposure does not evolve into atopic asthma.

Natural history of asthma

3.14   All longitudinal studies of asthma have, to date, been incomplete as regards their follow-up. Few are population based and have tended to follow high risk cases or those with fully established disease. Thus, studies have not taken into account the possibility that new cases arise at various stages in life. No single study has measured a comprehensive range of clinical, immunological and respiratory parameters.

3.15   The predominant pattern of illness in early life is characterised by episodic wheeze whereas, in adults, asthmatic patients may experience chronic persistent symptoms. The major prognostic factor for the evolution of intermittent disease to chronic disease of adult life is atopy. Thus, infants with eczema who also develop recurrent wheezing have a high probability of having persistent asthma in adult life. However, infants with wheezing only associated with intercurrent viral infection, rarely go on to develop atopic asthma. There may even be a mutually exclusive group of children who suffer wheeze with infection without atopy as compared with a group who have atopy and have a reduced frequency of viral infection induced wheeze in infancy. Whether this is cause or effect also requires elaboration. Asthma in mid-childhood is, however, associated with a higher probability of persistence through to adult life than has perhaps been appreciated. The studies from Melbourne, Australia, suggest that one major influence is cigarette smoking in adolescence and adult life, leading to a higher probability of persistence of asthma. By extrapolation, it might also be suggested that exposure to other pollutants would have a similar effect but this, again, requires further study.

3.16   Many studies of the adverse effects of air pollution in relation to asthma have focused on children. This, in part, may be explained by the investigators attempting to eliminate other confounding variables which might influence lung disease, including active smoking and occupational pollutant and allergen exposure. However, this ignores the fact that children are exposed to environmental tobacco smoke, sometimes in high concentrations, early in life. There is no doubt from many studies that such exposure increases morbidity from respiratory disease and particularly wheezing illnesses. It is also important to elaborate on the proportion of time that a child will spend exposed to particular factors. Children will spend twelve hours per day in their bedroom, five to six hours per day indoors at school, two to five hours in other rooms at home, leaving only a very small percentage of time spent out-of-doors with a potential for outdoor pollutant exposure. This illustrates that indoor exposures are likely to be immensely more important, merely because of the length of exposure. Unfortunately, no studies have been able to document duration of exposure in relation to any association and this will be very important for future studies.

Conclusions

3.17 Asthma is a common disease which, perhaps surprisingly, has proved difficult to define in a generally acceptable, conclusive way. In young children, cough and wheezing are commonly associated with respiratory infections, though in most cases such occurrences do not lead on to asthma in later life. It is possible that asthma is not a single disease entity and that the term "asthma" should be taken to include a number of more or less distinct syndromes.

3.18 A number of factors are known to induce asthma, and many more provoke attacks in those who have already developed problems (inciters). As might be expected, our understanding of factors which incite asthma is better developed than that of those which induce the condition.

3.19 Many of the points considered briefly in this chapter are dealt with in detail in other chapters of this report.

Chapter 4

Cellular and Biochemical Basis of Asthma

Introduction

4.1   There is no universally accepted definition of asthma; clinical features have been described elsewhere in this report (see Chapter 3) but include one or more of the following: cough, chest tightness or wheeze at night or in the morning or after exposure to various environmental stimuli.

4.2   Until recently, hyperresponsive contraction of airway smooth muscle was thought to be the primary cause of the acute airway obstruction seen in asthma.[1,2] Now, other characteristics are also considered important, including oedema and increased mucus secretion caused by inflammation. The emerging consensus is that chronic airway inflammation is a consistent component of asthma and that the severity of the disease reflects the degree of inflammation present even in those with mild or asymptomatic disease who require little or no treatment.[3] A minority of studies have failed to demonstrate significant inflammatory changes in the airways of asthmatic patients.

4.3   Atopy or allergy is a major risk factor for the development of asthma. Once sensitised to a specific allergen, the airways will respond to subsequent exposure with early (0–60 min) and later (2–24 h) phases of airway obstruction (EAR, LAR) and an increase in airways "non-specific" bronchial responsiveness that may persist for days or weeks after a single exposure. Induction of the early and/or late phase airway responses seen in allergic asthma, critically depends on the profile of inflammatory cells in the airway wall and varies between individuals. Thus, the early response of acute airway obstruction that occurs within a short time of inhalation of allergen is thought to depend largely on mast cell activation. In contrast, the late phase response, which usually occurs some hours later and is cytokine dependent, is more severe, does not occur in all subjects, is accompanied by an increase in bronchial hyperreactivity, is less responsive to bronchodilator therapy, involves influx of eosinophils and other inflammatory cells (see below) and the release of a different series of mediators from those seen in the acute response.[2,4-6]

Inflammatory cells and asthma

Mast cells

4.4   These cells are situated throughout the respiratory tract, between the bronchial epithelium and basement membrane, between the epithelial cells and sometimes near the surface of the epithelium close to the airway lumen. They are present in low numbers in non-asthmatic subjects at less than 1% of total cells recovered in bronchoalveolar lavage (BAL) but may increase several-fold in asthmatic subjects.[2] Although there is little evidence of increased numbers in the bronchial mucosa of asthmatic patients, their involvement in asthma, particularly in the early response, is most likely as a source of potent mediators that trigger subsequent events.[2,6]

4.5   In atopic individuals, binding of aeroallergen to mast cell bound IgE results in the release of inflammatory mediators including histamine, chemotactic factors for neutrophils and eosinophils, eicosanoid metabolites (see Figure 4.1) including leukotrienes (C_4, D_4 and E_4) and prostaglandins (eg PGD_2), and the neutral protease, tryptase, which can degrade neuropeptides and activate other mediators.[2,4,6] Mast cells contain the cytokines/lymphokines TNF–α, IL–4, 5 and 6 (see Figure 4.2).[7] Significantly, TNF–α expression is greater in mast cells from asthmatic patients compared with non-asthmatic subjects and correlates with the finding that BAL TNF–α levels are higher in symptomatic than in asymptomatic asthmatic patients.

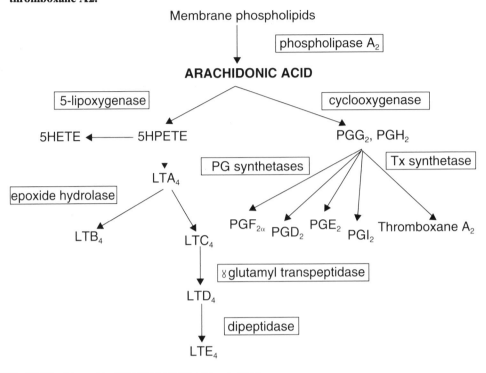

Figure 4.1 **Pathways involved in the synthesis of eicosanoids. Arachidonic acid is metabolised by 5-lipoxygenase to 5-hydroperoxyeicosatetraenoic acid (5HPETE) and hence to 5 hydroxyeicosatetraenoic acid (5HETE) or leukotriene (LT) A4 which in turn generates the chemotactic leukotriene LTB4, or the sulphidopeptide leukotrienes LTC4, D4 and E4. Metabolism by cyclooxygenase results in the formation of prostaglandin (PG) and thromboxane A2.**

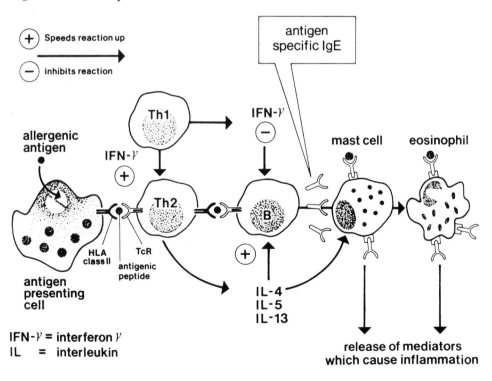

Figure 4.2 **Role of cytokines in asthma**

4.6   IL–4, which appears to be stored in mast cells under normal conditions, is more often detected in the secreted form in tissue taken from the airways of asthmatic patients[7] than in tissue taken from the airways of non-asthmatics. Release of stored TNF–α and IL–4 by mast cells may be important in activation of endothelial cells and subsequent inflammatory cell infiltration of the airway submucosa, precipitating the late asthmatic response. IL–4 is also important in the induction and maintenance of eosinophilia and in stimulating development of the Th2-helper T cell subtype (phenotype) which is also believed to orchestrate asthmatic inflammation. The presence of even low levels of IL–5 may also be important since this specifically stimulates eosinophil maturation and priming and prolongs their activity (see Figure 4.2). In non-atopic asthma, release of mast cell mediators, such as histamine, may be induced by cytokines or peptides released by other activated cells.[8]

Eosinophils

4.7   The eosinophil is believed to play an important role in the pathogenesis of asthma.[2,4,5,8] Eosinophil numbers are elevated in both BAL fluid and bronchial biopsies from asthmatic patients: their numbers relate broadly to the severity of the disease[2,5] and they are also the predominant cell in the allergen-induced late phase response to antigen challenge. They express low affinity receptors for IgE (Fc$_\varepsilon$R2, CD23) which may be an important signalling mechanism following antigen exposure. Eosinophils release a number of mediators[2,4,5,8] including leukotriene C$_4$ as well as platelet activating factor (PAF) (see Figures 4.1 and 4.2).

4.8   Eosinophils store at least four basic proteins:

Major basic protein (MBP) which is highly toxic to bronchial and alveolar epithelial cells as well as inducing hyperresponsiveness at high concentrations;

Eosinophil peroxidase which is also toxic to pulmonary epithelial cells, particularly in the presence of peroxide and halide ions;

Eosinophil cationic protein and eosinophil-derived neurotoxin which are also released during degranulation, are both toxic and have ribonuclease activity.

4.9   Although little is known about factors which stimulate eosinophil degranulation, IgA, and in particular secretory IgA, are well established activation mechanisms. However, the relevance of this or Fc$_\varepsilon$R2 to whether or not these cells can be triggered by air pollutants is not known. It has been unclear exactly why eosinophils but not neutrophils preferentially accumulate in the respiratory mucosa of asthmatic patients. A major difference between these cell types is the response to Th2 cytokines especially interleukin–3 (IL–3) and interleukin–5 (IL–5) which stimulates the relative proliferation and maturation of eosinophils, but not that of neutrophils (see Figure 4.2). In addition, IL–5 up-regulates eosinophil adhesion to endothelial cells and their response to chemokines such as IL–8. Together with granulocyte-macrophage-colony stimulating factor (GM-CSF), IL–5 also prolongs eosinophil survival. Thus, release of IL–5 by T cells and mast cells will favour eosinophil accumulation. Upregulation of the endothelial adhesion molecule VCAM–1 by IL–4 is also involved in eosinophil recruitment from the microvasculature through an interaction with its ligand VLA–4 on eosinophils.

Lymphocytes

4.10   Both B- and T-lymphocytes are normally present in the walls of the airways and on the pulmonary surface (see Figure 4.2).[2,4] B cells are responsible for antibody production and it has been suggested that elevated IgE levels seen in many asthmatic patients may reflect overproduction by B cells. There is no direct evidence that any single mechanism exists for overproduction of IgE by B cells in asthmatic patients, but defective regulation of B cells by T cells, involving such cytokines as IL–4 and IL–3 may be involved. More recent studies have concentrated on the role of T-lymphocytes in controlling the inflammatory events seen in asthma.[8–10,13]

4.11   T-lymphocytes in both the bronchial mucosa and BAL from atopic asthmatic patients have been shown to be in an heightened state of activation as a result of upregulation of the IL–2 receptor and the MHC class II molecule, HLA-DR which correlates with the number of eosinophils and degree of hyperresponsiveness of the airways to methacholine.[8–12] Activated CD4+ T cells, but not CD8+ T cells, can also be found in the peripheral blood and bronchial lumina of such subjects and fall in number following treatment in relation to reduction of severity of the disease.[8–10,13]

4.12   In mice, CD4+ T cells have been subdivided into two phenotypes. Th1 cells selectively secrete IL–2 and interferon-gamma (IFN–γ), while Th2 cells release cytokines encoded in a cluster on chromosome 5 including IL–4 and IL–5 (see Figure 4.2). Th1 cells appear to be involved with delayed-type hypersensitivity reactions, while Th2 cells are involved in allergic reactions (see Figure 4.2).

4.13   As in mice, there appear to be two subsets of CD4+ T helper cells in human lungs which can also be separated by their production of either IFN–γ or IL–5.[8–10,14] Both subsets produce GM-CSF and IL–3; INF–γ suppresses IgE production, while IL–4 induces it. Thus, subsets of T-lymphocytes that produce IL–5 and IL–4 will support eosinophil populations and stimulate IgE production and be associated with allergen-induced asthma. Subsets that produce IL–5 may primarily be involved in prolonging eosinophil action and may, therefore, be significant in non-allergic as well as allergic episodes of asthma. Proportionally high levels of T cell subsets producing INF–γ would not be expected to be associated with either eosinophil accumulation or IgE production and hence would be unlikely to be related to asthma. However, while this situation may apply to atopic asthma, much less is known about the role of T cells and their cytokines in late onset non-allergic asthma and occupational asthma due to low molecular weight reactive chemicals.

Alveolar macrophages

4.14   Alveolar macrophages are recovered in larger numbers than other cells in BAL fluid. They are also present, often as monocytes, throughout the tissues of airways and in the interstitium of the lung though these are not recovered by BAL. The observation that pulmonary bronchoalveolar macrophages appear to be important in the down-regulation of lymphocytes and the immune response (unlike macrophages from other organs) possibly reflects constant stimulation by aeroallergens and the advantages of reducing tissue damage that might be produced in response to these allergens. Alveolar macrophages possess low affinity Fc receptors for IgE (Fc$_\epsilon$R2, CD23) and may contribute to the allergic response,[2,15] since cross-linking of IgE or antigen on macrophages *in vitro* causes release of lysosomal enzymes, LTB$_4$, PGF$_{2\alpha}$ and large quantities of thromboxane A$_2$.

4.15   Macrophages also release histamine-releasing factors, histamine-release inhibitory factors and IL–1 which regulate mast cell histamine release.[16] These mediators may be relevant to the early asthmatic response.

4.16   In addition, release of potent chemotactic substances such as PAF, LTB$_4$ and low molecular weight peptides by activated macrophages may promote the inflammatory cell influx that occurs in the allergen-induced late asthmatic response.

4.17   The possibility that macrophages contribute to the pathophysiology of asthma is further supported by the finding that macrophage mediator release is inhibited by corticosteroids. However, as yet, no clear cut differences have been observed between bronchoalveolar macrophages from asthmatic compared with non-asthmatic individuals to suggest a major role for the macrophage in asthma. It is possible that interstitial macrophages or recently dendritic antigen presenting cells may have a pro-inflammatory role in asthma since their numbers are increased in the epithelium and submucosa in active disease.

Neutrophils

4.18   Whether or not the neutrophil is involved in asthma is controversial.[2,5,6] Most histopathological studies of lungs from patients who have died from asthma describe proportionally more eosinophils compared with neutrophils, regardless of the severity of the disease (with the exception of occupational asthma). Similarly, studies of cell profiles in sputum and BAL fluid from asthmatic patients indicate that the eosinophil, not the neutrophil, is the predominant inflammatory cell. Measurement of mediators in respiratory secretions from asthmatic patients also indicate that the eosinophils dominate. However, a recent study of seven cases of fatal asthma showed that whilst slow-onset fatal asthma involved more eosinophils and fewer neutrophils submucosally, in sudden-onset asthma the converse was true, suggesting a role for the neutrophil in sudden-onset fatal asthma.[17]

4.19   It is possible that neutrophils are important at the early stages or in acute (ie sudden-onset) asthma and hence are often not detected in tissue of BAL fluid samples obtained from chronically affected individuals. Animal models of allergic airway inflammation often implicate the neutrophil in the pathogenesis of asthma (see below). Neutrophils release many mediators of inflammation, including eicosanoid metabolites, prostaglandins and leukotriene B_4, as well as hydrolytic lysosomal enzymes and oxidants and apparently generate a histamine-releasing factor.[2,5]

Epithelium

4.20   The pulmonary epithelium plays a multifunctional role in providing a physicochemical barrier against inhaled toxic agents. The pseudostratified epithelium is maintained by a range of cell adhesion molecules. The basal cells are secured to the basement membrane via hemidesmosomes containing the integrin $\alpha_6\beta_4$. The columnar ciliated, goblet and Clara cells are attached to basal cells via desmosomes, complex structures containing at least three adhesion proteins, two glycoproteins and tonofilaments. Other adhesion proteins including intermediate junctions (uvomorulin) and CD44 provide additional adhesion mechanisms. Physically, the cells form tight cell-cell junctions that exclude organic and inorganic matter from the underlying tissue. The epithelium secretes a protective coat of mucus and surfactant which entraps foreign material which is then removed via the mucociliary escalator or is ingested by inflammatory, and other, cells. In addition, epithelial cells inactivate many inhaled toxins, and synthesise and secrete antioxidants (see below) and antiproteases (eg secretory leukoprotease inhibitor) which are important during increased oxidative or proteolytic stress (eg inhalation of ozone; release of proteases from inflammatory cells). Recent work shows that epithelium synthesises and secretes an array of potent mediators such as cytokines, lipid mediators (15-HETE, PGE_2), nitric oxide and endothelin which, with those of the migrating inflammatory cells, are likely to contribute to the chronic inflammatory response.[2,18,19] Other important mediators include modulators of smooth muscle tone (eg nitric oxide) and of neurogenic responses (eg neutral endopeptidase).[20,21]

4.21   It is therefore not surprising that damage to and shedding of the epithelium may play an important role in the pathology of asthma. Damage to the epithelium is probably in part the result of mediators released by inflammatory cells (eg major basic protein, oxidants and proteases) but may also result from the inhalation of cytotoxic agents. Loss of epithelial integrity may contribute to bronchial hyperresponsiveness by allowing easier access of irritants to intra-epithelial nerve endings, enhanced penetration of potential allergens/chemicals to underlying tissue and reduced control over smooth muscle tone due to loss of epithelial cell-derived relaxing factor and neutral endopeptidases.

4.22   Another important link between epithelial cell function and the pathophysiology of asthma is the possible synthesis and release of cytokines (including IL–1β, IL–6, GM-CSF, chemokines, RANTES and IL–8, and growth factors such as TNF-β, PGDF and bFGF).[18,19,22] These stimulate activation, proliferation, differentiation and migration of inflammatory cells and may affect fibrogenesis in airway wall modelling. Epithelial cells also produce leukotrienes (LTC_4 and LTD_4) and prostanoids (including PGE_2; see Figure 4.1).[2,18] The promotor regions for various epithelial derived cytokines and mediators contain the consensus sequence for the transcription factor NK-κB and are mostly downregulated by corticosteroids. The airway epithelium, therefore, serves an active role in asthma and may be of great importance in the maintenance of ongoing disease.

Muscles/nerves

4.23   Detailed accounts of the role of neuropeptides and activation of sensory, parasympathetic and sympathetic nerves on airway and vascular tone and on lung secretions are available.[21,23,24] Many neuropeptides have been localised to the respiratory tract and are likely to play a role in asthma (see Figures 4.3 and 4.4).

Figure 4.3 **Axon reflex mechanisms in asthma. Damage to airway epithelium in asthma exposes unmyelinated nerve endings which may be triggered by mediators such as bradykinin, resulting in resistance of sensory neuropeptides such as substance P (SP), neurokinins (NK) and calcitonin gene-related peptide (CGRP). Together these neuropeptides produce the pathology of asthma.**

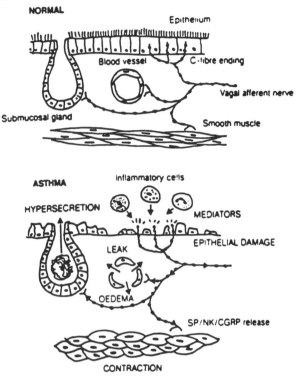

Figure 4.4 **Airway inflammation and VIP co-transmission. Normally, vasoactive intestinal peptide (VIP) and peptide histidine methionine (PHM) counteract cholinergic contraction of airway smooth muscle. In asthma, inflammatory cells may release enzymes which destroy these peptides, thus enhancing cholinergic nerve effects resulting in exaggerated bronchoconstriction.**

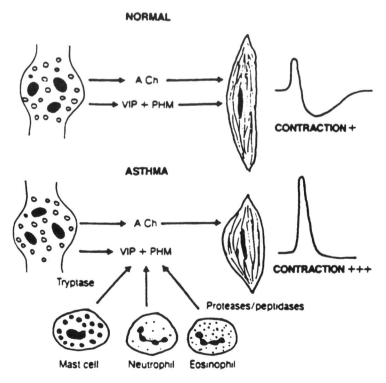

The so called "sensory neuropeptides" have attracted attention since they induce many of the pathophysiological features of asthma. For example, neurokinin A is a potent bronchoconstrictor, while substance P and calcitonin gene-related peptide (CGRP) are vasodilators; substance P also induces hypersecretion. The activity of such peptides is dependent not only on their release but also on the rate at which they are metabolised (see Figure 4.3). Neutral endopeptidase inactivates these peptides; consequently reduced neutral endopeptidase activity may lead to enhanced bronchoconstriction. In contrast, vasoactive intestinal peptide (VIP), one of the most abundant neuropeptides in the lung, localised mainly to the cholinergic nerves, is a potent relaxant of airway vascular smooth muscle and may regulate bronchial smooth muscle tone (see Figure 4.4). VIP is susceptible to proteolytic inactivation by mast cell tryptase, which is elevated in asthmatic patients and which prevents airway relaxation induced by VIP (see Figure 4.4).

4.24 It is unclear how these neuropeptides are activated or depleted in asthma; however, epithelial shedding, exposure of nerve endings and their activation by exogenous agents, reduction or increase in mucosal proteases that control neuropeptide activity due to epithelial damage or inflammatory cell influx and release of other cellular mediators are all possible factors (see Figure 4.3). It is likely that there is a complex interplay between neuropeptides which results in the finely modulated airway responses of health and which, if disturbed, may play an important part in asthma.

Role of antioxidants

4.25 Recent reviews summarise much of our present knowledge of the antioxidant potential of respiratory secretions.[25,26] The major antioxidants in respiratory secretions are uric acid, reduced glutathione, mucin, protein (mainly albumin) and ascorbic acid (see Table 4.1a).

Table 4.1a **Approximate values for the concentrations of nonenzymatic antioxidants in human plasma as compared to peripheral RTLF [epithelial lining fluid (ELF)] (from reference 25)**

Antioxidant	Plasma, μM	ELF, μM
Ascorbic acid	40	100
Glutathione (GSH)	1.5	100
Uric acid	300	90
Bilirubin	10	–
α-tocopherol	25	2.5
β-carotene	0.4	–
Ubiquinol-10	0.6	–
Albumin-SH	500	70

Metal binding proteins such as lactoferrin, and enzymes, such as catalase and extracellular superoxide dismutase, have also been detected and may be important.[25-27] Nasal secretions are enriched with uric acid which is an effective scavenger of ozone (and NO_2, SO_2).[28] Uric acid also contributes a significant proportion of the antioxidant capacity of epithelial lining fluid (ELF) from the proximal and distal airways.[25,29] ELF additionally contains a large amount of glutathione, which scavenges $\cdot OH$, H_2O_2 and HOCl (produced by activated phagocytes) and other oxidative molecules.[25,26]

4.26 Levels of glutathione are significantly elevated in respiratory secretions in asthmatic patients.[30] Ascorbic acid is present at effective levels in the respiratory lining fluid of some but not all individuals. This might reflect dietary intake. Mucin consists of core glycoproteins which are rich in -SH and S-S groups which react with oxidative molecules. Stimulation of mucus production follows inhalation of some toxic compounds and could, therefore, be protective. Vitamin E is lipid soluble and is most likely associated with membranes; it is present in ELF, probably within micelles, and may protect against peroxidative injury.

4.27 The source of antioxidants varies from compound to compound. Ascorbic acid, but not uric acid, is more concentrated in respiratory tract secretions than in plasma, suggesting an active secretory process. This is not yet understood.[25] Glutathione and mucin are synthesised by epithelial cells and levels in ELF appear to be related to the state of epithelial integrity and activation.[25,26] Macrophages also contain glutathione, release of which depends on their state of activation. Proportionally, uric acid is an important antioxidant especially in nasal secretions but also in ELF.[25,29,30] It is suggested that uric acid is actively taken up from the plasma and released by airway gland cells. Most of the protein in ELF is albumin, formed by transudation from the plasma. Albumin, together with other sulphur-containing proteins, act as antioxidants.

4.28   The relative importance of the different antioxidants in protecting the lung from oxidative stress is not known and is the subject of intense research. Surprisingly, little is known about their production, utilisation or turnover and how these processes are regulated in the lung. Preliminary studies of the effects of ozone on human nasal and bronchoalveolar lavage fluids indicate that uric acid is rapidly consumed in direct relation to its original concentration prior to exposure.[29,30] These types of study are in their infancy but will be critical in assessing the relative protection afforded by, and the significance of low levels of, individual antioxidants within the lung. Comparative studies show that there are significant differences between rodent and human ELF antioxidant profiles and this will need to be considered in interpreting the results of studies on animals (see Table 4.1b).[25]

Table 4.1b   **Comparison of estimated levels of certain antioxidants in ELF from different mammals (from reference 25)**

Antioxidant	Guinea Pig	Rat	Human
Ascorbic acid	600 μM	2000 μM	160 μM
Glutathione	220 μM	130 μM	165 μM
Uric acid	53 μM	46 μM	129 μM
α-tocopherol	0.8 μM	0.6 μM	2.5 μM
Protein-SH	150 μM	75 μM	60 μM

The pathology of asthma

4.29   The pathology of asthma may be studied by post-mortem examination of lungs from asthmatic patients dying of status asthmaticus or from unrelated causes, by examination of bronchial biopsies and by examination of sputum or of fluid collected at bronchoalveolar lavage.

Post-mortem appearances: macroscopic

4.30   The lungs of patients who have died during a severe attack of asthma are hyperinflated, though small areas of collapse and bronchiectasis, usually at the periphery of the upper lobes may be found. The parenchyma of the lung is well preserved and emphysema is not characteristic. Thickening of the walls of large and small airways may be seen and airways of 0.2–1.0 cm diameter are often blocked with mucous plugs.[31]

4.31   In asthmatic patients dying from other causes, some over-distension is often seen, and the airways may be lined with a layer of exudate.[31]

Histological appearances

4.32   The airways of asthmatic patients dying of status asthmaticus contain a mixture of exudate, mucus and cellular debris. Cells in the debris include eosinophils, exfoliated columnar epithelial cells in various stages of generation and squamous metaplastic cells. The exudate may be found as far as the distal airways and if blockage of small airways has occurred, distal collapse may be present.

4.33   The bronchial mucous membrane is widely damaged with loss of ciliated cells and intraepithelial nerves are exposed. In areas of severe damage, stripping of epithelial cells may be almost complete and in some areas only basal cells remain attached to the basement membrane. Basal cells not only represent the reserve cells of the epithelium, they probably also play an important role in anchoring the other epithelial elements, particularly the columnar ciliated cells, to the basement membrane. Between attacks goblet cell hyperplasia and focal squamous metaplasia are evident. Mast cells may be seen amongst the epithelial cells. Increased penetration of the epithelium by mast cells seems to be characteristic even when the general epithelial structure is fairly well preserved. In attacks they appear to be reduced because they have undergone degranulation.

4.34 The epithelial basement membrane region of the asthmatic airway is thickened: in some cases a marked increase in thickness from a normal of $7 \mu m$ to about $30 \mu m$ may be seen. Lesser degrees are often seen in non-asthmatic patients, so as a feature it is evidently not specific to asthma. Electron microscopy has shown that the thickening involves the deepest layer of the basement membrane, the collagen and fibronectin-rich lamina reticularis rather than the true basement membrane. While the normal basement membrane contains predominantly types IV and VII collagen together with heparan sulphate, in asthma, the increased density of the lamina reticularis is due to deposition of interstitial collagens (types I and III) secreted by a network of subepithelial myofibroblasts whose number is increased. Immunoglobulins are found in the basement membrane but they are also present in many non-asthmatic patients.

4.35 The submucosa of the airways of the asthmatic patient is oedematous and contains dilated blood vessels, many eosinophils and lymphocytes. Mast cells are also seen.

4.36 The glands of the submucosa of the segmental airways are enlarged in asthma but not as markedly as in chronic bronchitis. Dunnill has pointed out that in chronic bronchitis there is evidence of hyperplasia of these glands; in asthma, enlargement by hypertrophy seems to be more common and involves both serous and mucous acini while in chronic bronchitis only mucous acini are affected.[31]

4.37 The amount of smooth muscle in the wall of the airways is increased in asthma due to hyperplasia.[32-35] Modelling of the mechanics of the airway wall has shown that for a given degree of muscle shortening the increase in airway resistance will be related to the thickness of the wall.[35,36] This has been suggested as an explanation for the enhanced bronchial hyperresponsiveness characteristic of asthma. The role of changes in the structure of the basement membrane in airway collapse has recently been studied.[37] It has been suggested that abnormal mechanical functioning of the basement membrane could predispose to collapse of the airway despite comparatively normal functioning of the smooth muscle layer.

Biopsy studies

4.38 Many of the changes described above have been observed in biopsy specimens obtained between attacks.[38,39] Thickening of the basement membrane, mast cell infiltration and degranulation, infiltration of eosinophils into the epithelium and shedding of epithelial cells have all been described. It is suggested that these changes are continuous in asthmatic patients and are enhanced during exacerbations of the disease. The exposure of intraepithelial nerves through damage to the epithelium may also contribute to the hypersensitivity to a range of stimuli seen in asthmatic patients.

Sputum

4.39 Clumps of epithelial cells (Creola bodies), twisted casts of small airways (Curschmann's spirals) and Charcot Leyden crystals derived from granules of eosinophils are the hallmarks of sputum produced by asthmatic patients.[40]

Lavage fluid

4.40 Increased numbers of ciliated cells, eosinophils, monocytes and macrophages have been found in lavage fluid of asthmatic patients. It has also been demonstrated that the recovery of epithelial cells in lavage fluid, a measure of damage to the epithelium, may be related to the airway responsiveness to inhaled histamine. Damage to ciliated epithelial cells would also be likely to impair the clearance of mucus from the airways.

Possible mechanisms of effect of air pollution in asthmatic individuals

4.41 Air pollutants may affect asthmatic patients via a number of mechanisms: a number of which are considered below. This section has been followed by a detailed description of the animal models which have been developed to investigate these hypotheses.

(a) Is there a direct effect of pollutant on the airway epithelium?

4.42   At high concentrations sufficient to overcome the airway's natural defence mechanisms, air pollutants may cause epithelial cell death and denudation of the airway. This would expose underlying nerves and tissue to the direct effects of air pollutant as well as aeroallergens, possibly causing initial sensitisation, increased sensitivity to allergens or exacerbations of asthmatic responses.

4.43   An alternative mechanism of action of air pollutants on epithelial cells might involve epithelial cell cytokines which play a significant role in the control and induction of airway inflammation.

(b) Does exposure to air pollutants alter airway defence mechanisms, eg antioxidants and antiproteases?

4.44   Reduced activity of these defences could lead to increased oxidative or proteolytic damage, "leakiness" of the epithelium and greater accessibility of subepithelial tissues to allergens and other agents that precipitate asthma attacks.

(c) How might air pollutants interact with luminal inflammatory cells?

4.45   Pollutants may stimulate release of mediators of inflammation, increasing airway inflammation and precipitating bronchoconstriction. For example, O_3, NO_2 and transition metals in inhaled particles have the capacity to generate reactive oxygen intermediates in epithelial cells. These in turn are able to activate the transcription factor NF-κB, which, on passing into the nucleus and binding to a consensus sequence in the regulatory elements of specific genes, leads to the upregulation of cytokines (especially chemokines and TNF-α) and cell adhesion molecules (eg, E-selectin, ICAM–1 and VCAM–1). Macrophages may also be significant in this process, releasing chemoattractants for other inflammatory cells as well as lysosomal enzymes, nitric oxide and products of arachidonic acid metabolism.

4.46   In addition, pollutants might act on the effector side of the asthmatic response by a number of mechanisms including:

> lowering of stimulatory threshold, of, for example, smooth muscle;
>
> increasing likelihood of sensitisation;
>
> increasing response to allergen.

These have been studied in volunteers, and are dealt with in Chapter 5.

Studies in animal models

4.47   Ideally, these models might be divided into two categories:

a.   those involving species which develop spontaneous episodes of bronchoconstriction;

b.   those in which bronchoconstriction may be induced under experimental circumstances, such as following challenge with antigen, pharmacologic agents or inhaled irritants.

4.48   Unfortunately, there are few natural examples of animals which cough or wheeze spontaneously.[41–43] The exceptions are horses (ponies) and primates. Horses develop "heaves" which is associated with allergy and airway hyperreactivity and involves episodes of acute airway obstruction. The immunopathology, however, is more like extrinsic allergic alveolitis than asthma since it also involves the alveoli. Primates may become sensitised to *Ascaris suum* (naturally or experimentally) and exhibit spontaneous episodes of bronchospasm. The immunopathology reflects eosinophil and mast cell mediated inflammation but does not have the appearance of the chronic inflammatory response so characteristic of asthma.[44]

4.49   The animals most commonly used in the study of asthma are guinea pigs, rats, mice, rabbits, dogs and sheep.[41–43,45,46] No animal model mimics fully the range of responses seen in man and selection of an animal model usually depends on the aspect of asthma under study. The situation inevitably becomes more complex when studying possible additive or synergistic effects of air pollutants. A common situation in man is likely to be exposure to air pollutants on a background of chronic asthma. In the absence of a chronic model of asthma satisfactory prediction of likely response from studies in animal models is unlikely. Other disadvantages of using animal models to study asthma include differences from man in airway anatomy, inflammatory cell profile and mode and degree of physiological response.

Animal models of asthma

Ideal model of asthma

4.50 For comparison with man, an ideal model of asthma would include the following:

(1) Chronicity and intermittent "spontaneous" relapses in severity.

(2) Pathological features of chronic inflammation involving mast cells, eosinophils and T cells, epithelial damage and airway wall remodelling, eg, increased interstitial collagens in the lamina reticularis of the basement membrane.

(3) Non-specific airway hyperresponsiveness *in vivo*, to include hypersensitivity (low threshold responsiveness) and hyperreactivity (steep slope of dose-response curve).

(4) The absence of a plateau response to a bronchoconstrictive stimulus.

(5) Efficacy of therapeutic modalities similar to that observed in human asthma.

(6) A late response to allergen challenge.

(7) Bronchoconstrictor responses to beta adrenergic blockers.

(8) Increased lung volumes during a response to allergen challenge.

Allergic models: immunologically induced

4.51 These models most often involve prior sensitisation of experimental animals to foreign proteins: a proportion of the animals develop hyperresponsive airways. As mentioned earlier, some of the larger animals do not need to be actively sensitised as they are naturally hyperresponsive to some allergens (see Table 4.2).

Table 4.2 **Advantages and disadvantages of various animal models of asthma (Taken from reference 43)**

Advantages	Disadvantages
Mouse	
Small, inexpensive animal	Poorly developed airway smooth muscle
Numerous inbred strains	Responds poorly to histamine
Availability of species-specific reagents	
Demonstrates AHR to carbachol	
Rat	
Small, inexpensive animal	Injection of antigen required for sensitisation
IgE is major class of anaphylactic antibody	Adjuvants (alum, *Bordetella pertussis*) required
Demonstrates EAR and LAR	for sensitisation
Responds to methacholine	Responds poorly to histamine
Responds to cromolyn	Tracheal smooth muscle responds poorly to
	peptide-leukotrienes
Guinea-pig	
Small, docile animal	Few inbred strains exist
Relatively inexpensive	Few species-specific reagents
Sensitisation possible via inhalation route	IgG$_1$ is the major class of anaphylactic antibody
Lung is major shock organ	Not sensitive to cromolyn
Demonstrates EAR and LAR	
Neutrophil influx to lung following LAR	
Eosinophilic inflammation during LAR	
AHR following airway hypersensitivity response	
Tracheal smooth muscle responds to histamine	
Rabbit	
Lung is anaphylactic target organ	Neonatal immunisation required for LAR
IgE is major class of anaphylactic antibody	
Demonstrates EAR, LAR	

Advantages	Disadvantages
Dog	
IgE is major class of anaphylactic antibody	Individual variability in responsiveness to *A.*
Natively sensitised to *Ascaris suum*	*suum* LAR not demonstrated
Demonstrates AHR	
Basenji greyhound dog	
Persistent AHR	AHR not associated with clinical disease
Very responsive to methacholine	
Pony	
Native sensitisation to "barn" environment	Large animal, costly
Develops AHR (only during clinical disease)	
Recurrent exacerbations of clinical disease	
Sheep	
Natively sensitive to *A. suum*	Large animal, costly
Demonstrates EAR and LAR	
Demonstrates AHR to carbachol	
Nonhuman primate	
Consistent response to antigen	Costly species
Natively sensitised (*A. suum*)	
Allergic responsiveness persists for years	
IgE is major class of anaphylactic antibody	

AHR: airway hyperresponsiveness; EAR: early airway response; LAR: late airway response; IgE: immunoglobulin E; IgG: immunoglobulin G

Guinea pig

4.52   The guinea pig is one of the most commonly used models of asthma as sensitisation is easily achieved with a single low dose of ovalbumin (eg, 1 mg/kg, subcutaneously) and the lung is the target organ in the anaphylactic reaction induced on subsequent challenge.[45] In addition, guinea pigs develop an early and late onset response and have well developed airway smooth muscle that contracts extensively and rapidly following challenge.[41,42,45] However, humans do not have such well developed airway smooth muscle. Furthermore, unlike man, bronchoconstriction in the guinea pig is largely histamine-dependent and cannot be inhibited by corticosteroids.[41,45] Another difference is the immunological response which, at high doses of allergen (to induce the greatest airway response), is IgG$_1$ mediated, whereas the commonest atopic form of human asthma is related to an IgE response. However, in more recent studies where guinea-pigs were aerosol-sensitised (rather than subcutaneously) the IgG response was minimal, with an inflammatory response in the early and late phases which seem to resemble those in man; that is an early neutrophilia (within 3–4 hours) and a late eosinophilia (18–24 hours) which is related to a late phase response to allergen challenge.[47,48]

Rat

4.53   Sensitisation of rats is more difficult to achieve requiring immunisation with higher doses of allergen with adjuvants such as alum and *Bordetella pertussis*.[41-43] Nevertheless, the Brown Norway rat has been much used as a model for asthma since its immunological response is, like man's, primarily IgE-mediated.[43] Following challenge with aerosolised ovalbumin, the animals exhibit an early and late onset response and the number and type of inflammatory cells and activated T cells present resemble those seen in man.[49] An important difference between the guinea pig and rat is that bronchoconstriction in the rat responds to both corticosteroids and sodium cromoglycate resembling that of man.[41] The major difference so far between the human and rat is that the airway smooth muscle response to antigen is primarily mediated by serotonin (5HT), although, like man, the parenchymal smooth muscle response can be mediated by LTD$_4$.[41]

Mouse

4.54　Until recently, murine models of asthma were rare. With the realisation of the complex role of lymphocytes, lymphokines and cytokines in human asthma and the fact that mice are the species which have been most exhaustively studied and characterised at the immunological level, the use of murine models of asthma is increasing.[43,45] A drawback to the previously limited use of mice to study asthma has been measurement of pulmonary function. However, methods are now available to make appropriate measurements of murine lung function and evidence suggests that the mouse is a good model.[50] Like the rat, one of the murine models involves prior sensitisation with ovalbumin and challenge with aerosolised ovalbumin. Mice exhibit bronchial hyperresponsiveness, airway inflammation and synthesise specific IgE.[43,45,50] In particular, the inflammatory response with respect to eosinophil infiltration and the T- helper cell subtypes resembles that seen in man.

Rabbit

4.55　Rabbits will develop both early and late phase responses following active sensitisation with antigen, eg house dust mite, *Dermatophagoides pteronyssinus*. This response usually requires the use of adjuvant. When sensitisation to *Alternaria tenuis* begins during the neonatal period (ie within 24 hours of birth) for up to three months, it is possible to induce a specific IgE response, the rabbits developing both early and late phase pulmonary responses following challenge with aerosolised antigen and histamine.[43,51] These changes in airway hyperresponsiveness were subdued in the presence of IgG to *Alternaria*, lasted for up to 12 months in a sub-population of animals and appeared to be unrelated to inflammatory cell infiltration or bronchial smooth muscle cell responsiveness.[43,52] However, other rabbit models of airway hyperresponsiveness have been associated with airway inflammation during the late phase response.[53]

Dog

4.56　Dogs provide probably the most readily available large animal model of asthma.[43,46] Dogs sensitised with aerosolised *Ascaris suum* or ragweed develop IgE antibodies and airway inflammation. Early studies suggested that the inflammation was primarily neutrophilic rather than eosinophilic at 24 hours, although resident eosinophils appear to be significant in the subsequent airway response to aerosolised antigen.[54] Mongrel dogs do not exhibit a late phase response to challenge although hyperresponsiveness is persistent in Basenji greyhounds and they also exhibit a form of allergic eczema.[43]

Sheep

4.57　Sheep have also been studied extensively following natural or active sensitisation with *Ascaris suum* and aerosol challenge and have the advantage that a proportion develop a late response.[42,43,46,55] IgE levels are raised in sensitised animals and the late onset response involves an influx of eosinophils and neutrophils which responds to glucocorticoids.[56] The allergic sheep model has been favourably compared with allergic asthma in man with respect to airway responsiveness, the profile of airway inflammation and the response to pharmacological intervention. However, the immunopathology lacks the chronic cellular elements and features of airway wall remodelling characteristic of asthma.

Primates

4.58　Non-human primates can also be actively sensitised with *Ascaris suum* and challenged with aerosol.[42,43] In some animals, sensitisation may have been acquired without experimental intervention. Allergic Cynomolgus and squirrel monkeys develop both early and late onset responses.[57,58] Because a proportion of non-human primates develop a dual response to antigen and with the obvious advantage that their cellular and immunological background is close to man, they are often regarded as representing the best model of human allergic asthma. Interestingly, a recent study in Cynomolgus monkeys that were naturally sensitised to *Ascaris suum* showed that animals producing both early- and late-phase response to antigen inhalation challenge had higher levels of eosinophils in BAL prior to antigen challenge and that the late phase response was related to the increase in BAL neutrophils.[57] In those showing only a single response, eosinophils were elevated only after challenge, and the increase in BAL neutrophils was lower than that in dual responders. Furthermore, chronic treatment of the dual responders (7 days) with dexamethasone prior to antigen challenge caused a significant reduction in BAL eosinophils as well as significantly blocking the late bronchoconstrictor response and associated eosinophil and neutrophil influx.[57]

Relationship between allergy, air pollution and asthma

4.59   In man, the asthmatic response is complex and involves a large range of mechanisms or pathophysiological processes. In the animal models described above, allergy is the underlying trigger of hyperresponsiveness, inflammation and bronchoconstriction that occur following subsequent challenge. One of the features which remains unclear in man is the significance of atopy to the bronchoconstrictor response which may be precipitated by exposure to air pollution. Some individuals exhibit airway hyperreactivity in the absence of any obvious underlying tissue pathology and it may be that such individuals may be more susceptible to airborne pollutants. Studies in animal models are often difficult to assess or extrapolate to man in the context of existing atopy/asthma. In addition, there are forms of asthma with features of an eosinophil mediated inflammation but which are not associated with atopy. Late onset and aspirin sensitive asthma are examples of this.

Ozone

4.60   A previous Department of Health report on ozone has concentrated on the damaging effect of ozone on lung structure, specific interaction with cells and acellular components of lung tissue, defence mechanisms in the lung against ozone exposure and mechanisms of repair.[26] Many of the studies described in that report address the long term effects of ozone; the following account will concentrate on studies pertinent to the role of ozone in asthma. As mentioned in other sections of the report, acute exposure to ozone induces airway hyperreactivity and inflammation in man and, not surprisingly, it has been suggested as a possible explanation for the effects of air pollutants on asthmatics.

4.61   Guinea pigs, rats, cats, dogs and monkeys develop airway hyperreactivity following acute exposure to ozone in doses ranging from 1 to 3 ppm (2 to 6 mg/m³). A variety of stimuli have been used to induce subsequent bronchoconstriction, including histamine, acetylcholine, platelet activating factor, substance P, 5-HT and citric acid. These are generally more effective when given by inhalation.

4.62   In some models, ozone also induces increased airway permeability and inflammatory cell infiltration. The exact relationship between ozone exposure and the development of epithelial permeability, airway inflammation and airway hyperreactivity remains unclear.

4.63   Early studies in mongrel dogs showed a strong relationship between the increase in the number of neutrophils in the airway epithelium and airspaces and the degree of airway hyperresponsiveness (to acetylcholine) 1 hour after exposure to 2–3 ppm (4–6 mg/m³) ozone for 2 hours;[59–61] animals which did not develop airway hyperreactivity did not show an increase in neutrophil numbers.[59,60] Where the number of neutrophils recovered by BAL increased, epithelial cell shedding was also observed.[60] Furthermore, prior depletion of circulating neutrophils inhibited bronchial hyperresponsiveness.[62] The effects of ozone disappeared 1 week after exposure. Subsequent studies in the same model using either the same exposure protocol or reducing the ozone exposure time (to 20 minutes) suggested that ozone-induced airway hyperreactivity might be due to generation of oxygen radicals by neutrophils and generation of arachidonic acid metabolites which mediate smooth muscle contraction.

4.64   Treatment with antioxidants (desferoxamine mesylate which chelates iron to prevent generation of iron-dependent hydroxyl radicals and allopurinol which inhibits xanthine oxidase)[63] or indomethacin (prostaglandin synthetase [cyclooxygenase] inhibitor)[61] prevented increased airway hyperreactivity but not neutrophil influx. Interestingly, a later study showed that although inhalation of the corticosteroid, budesonide (2.66 mg/day, 7 days prior to 3 ppm (6 mg/m³) ozone, 30 minutes), caused a significant reduction in ozone-induced neutrophil and eosinophil influx as well as airway narrowing, it did not prevent airway hyperreactivity to acetylcholine challenge.[64] This later study supports some of the rodent studies (see below) that suggest that ozone-induced airway hyperreactivity can be independent of inflammatory cell influx and their mediators.

4.65   A number of studies in guinea pigs and rats also suggest that ozone-induced airway hyperreactivity is independent of neutrophil influx and possibly independent of an increase in vascular permeability. In these studies, animals were exposed to between 1 and 4 ppm (2 and 8 mg/m³) ozone for from 15 minutes to 2.25 hours.[65-68] Hyperreactivity was usually measured within 2 hours of ozone exposure, for up to 4 days depending on the design of the study. In contrast to many of the studies on dogs, most of the rodent data suggest that hyperreactivity occurs in advance of neutrophil influx.[65-67] The sequence of events after acute ozone exposure appears to be increased hyperreactivity within 30 minutes of exposure, followed by increased protein permeability over the next 4 hours then an influx of neutrophils between 4 and 12 hours later.[65,67] The early increase in permeability is likely to be due to epithelial rather than vascular damage.

4.66   Studies on tracheal capillary and epithelial permeability using horse radish peroxidase as a marker of permeability showed that, following 3 ppm (6 mg/m³) ozone for 30 minutes, the epithelium became leaky within 20 minutes which coincided with induction of the greatest airway responsiveness to methacholine which then returned to normal within 2 hours.[69] In contrast, tracheal capillary leak was maximal after 1 hour and returned to control levels within 4 hours.

4.67   Some of the most thorough studies of the sequence of cellular events after acute ozone exposure involve rats, but unfortunately have not included concomitant studies on hyperreactivity. However, Evans *et al* demonstrated hyperreactivity in neutrophil-depleted Long Evans rats immediately following 4 ppm (8 mg/m³) ozone for 2 hours, which additionally was not related to tracheal vascular permeability, suggesting ozone-induced hyperresponsiveness is independent of inflammation.[66] Other studies in rats, often using lower doses of ozone, suggest that the most likely pathological sequence is damage to the epithelium, vascular damage, plasma and cellular protein leak followed by an inflammatory cell influx.[70,71]

4.68   Although epithelial damage may be striking at higher exposure levels and involve cellular cytotoxicity and sloughing off of epithelium, low exposure levels may involve more subtle changes such as damage to the tight junctions between the cells. As in the guinea pig model, tracheal permeability to ^{99m}Tc-diethylene triaminopentaacetate (DTPA) in rats was increased immediately following exposure to 0.8 ppm (1.6 mg/m³) ozone for 3 hours.[70] This increase in permeability peaked at 8 hours. Other similar studies by the same group shows that neutrophil influx occurs much later, peaking between 8 and 12 hours.[71] Studies in neutropenic rats suggested that the early ozone-induced leak was independent of neutrophil influx but that protein leak was exacerbated by inflammatory cells and their products, such as leukotrienes.[72]

4.69   Longer exposures (4–24 hours) of normal and neutropenic rats to 1 ppm (2 mg/m³) ozone essentially results in the same sequence of events—early epithelial damage and protein leak followed by neutrophil influx.[73,74] It was concluded that neutrophils did not contribute to epithelial damage in these studies. However, in the latter studies, the investigators were primarily interested in the necrosis of epithelial cells and inflammatory cell influx within, and distal to, the terminal bronchioles since these are the areas in which the most lasting morphological changes were expected to be after inhalation of ozone. Detailed study of the histopathology of the airway mucosa of guinea pigs for up to 4 days after acute exposure to 3 ppm (6 mg/m³) ozone for 2 hours showed that ozone had a similar effect on large airways. The significant injury to ciliated columnar and mucosal goblet cells coincided with the increase in mast cells (rather than neutrophils) and hyperreactivity described above.[65]

4.70 Chronic exposure (6 hours/day, 5 days/week) of Cynomolgus monkeys to 1 ppm (2 mg/m³) ozone did not result in increased airway hyperreactivity; this difference between studies may reflect the lower concentration of ozone or, alternatively, induction of ozone tolerance after 12 weeks exposure in the monkey model.[75]

4.71 A significant proportion of these investigations have concentrated on the involvement of the neutrophil in the very early effects of ozone on the lung. This reflects the consistent finding of neutrophil influx following ozone exposure. However, some of these studies suggest that mast cell and eosinophil numbers may also change following ozone exposure. As both these cells are thought to have a central role in human allergic asthma, there are important observations. In guinea pigs, Murlas and Roum[65] found increased tracheal mucosal mast cells within 2 hours of ozone exposure (3 ppm, 6 mg/m³; 2 hours) which remained elevated for 24 hours and occurred in advance of neutrophil influx. Similar observations were made in sheep exposed to 0.5 ppm (1.0 mg/m³) ozone for 2 hours.[76] In a study in mice which were either inherently deficient in, or which had been depleted of, mast cells, ozone did not induce neutrophil influx.[77] Since mast cells contain stored mediators both of airway hyperreactivity and of neutrophil and eosinophil chemotaxis, their presence/ increase following ozone exposure could be important. Although in the study of Murlas and Roum[65] guinea pig mucosal eosinophil numbers dropped over the first 24 hours then returned to normal 48 hours after ozone exposure, Tan and Bethel[78] found an overall increase in both BAL and mucosal eosinophils 4–5 hours after the same ozone exposure protocol, supporting the possibility that mediators released by resident mast cells (mast cell numbers did not increase at this time interval) may be involved.

4.72 Results from these non-allergic animal models suggest that inhalation of ozone at concentrations between 0.8 and 3 ppm (1.6 and 6 mg/m³) for up to 4 hours induces hyperreactivity that is initially independent of infiltrating inflammatory cells. The epithelial integrity throughout the airway is damaged which presumably increases the accessibility of underlying cells/tissue/molecules to ozone or, more likely, reactive species generated by the reaction of ozone with surface components. Subsequent vascular damage allows bi-directional protein leak which is followed by neutrophil and possibly eosinophil influx. It is suggested that the increase in neutrophils and eosinophils reflects early stimulation of resident mast cells (which may also increase in number) by ozone to release stored pro-inflammatory mediators.

4.73 Furthermore, the early ozone-induced epithelial cell damage/activation may also stimulate release of pro-inflammatory mediators. Subsequent release of mediators by sequestered neutrophils and eosinophils exacerbates the degree of ozone-induced inflammation.

4.74 The exact biochemical/molecular mechanisms involved in the hyperreactive response to ozone is unclear. As mentioned above, histamine release by mast cells may be significant in the guinea pig since bronchoconstriction is largely histamine dependent. Other studies in the guinea pig, using similar ozone exposure regimes, have addressed the role of peptide mediators in ozone-induced hyperreactivity. For example, stimulation of afferent nerves causes the release of tachykinins which are small peptides that produce tachypnoea, increased tracheal permeability and may mediate increased airway hyperresponsiveness.[21,24]

4.75 Capsaicin depletes sensory nerve terminals of tachykinins; prior treatment of guinea pigs with capsaicin partially inhibited ozone-induced hyperreactivity (but not increased protein permeability) suggesting at least a partial role for these peptides in ozone-induced airway hyperreactivity.[79] Neutral endopeptidase (NEP) is located on the external surface of airway mucosal cells and proteolytically inactivates substance P, which in turn causes hyperreactivity. Ozone-induced substance P-induced hyperreactivity is inhibited by prior ozone exposure but can be reversed by aerosolised NEP.[68] Furthermore, measurement of tracheal homogenate NEP activity showed that it was inhibited by ozone; its location on the external cell surface may therefore render it highly susceptible to inactivation by ozone. These studies indicate that one mechanism of action of ozone in induction of hyperreactivity may be via direct interaction with molecules that modulate activation of sensory nerve fibres.

Oxides of nitrogen

4.76 The potential of nitrogen oxides to interact with cellular and acellular components of lung tissue has been discussed extensively in a previous report.[80] Like ozone, it is likely that oxidative modification of proteins and/or formation of lipid peroxides is an early event. However, there is little information on the exact molecular sites of these actions *in vivo* and hence no understanding of the biochemical processes that might precipitate subsequent cellular and physiological airway responses.

4.77 Although it has been possible to induce hyperreactivity to NO_2 in animal models, high "irritant" doses (eg, >100 ppm, 188 mg/m^3) have usually been used and often chronic or sub-chronic exposures (ie, days or weeks rather than hours) have been necessary. Unlike the studies with ozone, the results are inconsistent and the sequence of early and cellular and physiological events in the airways following NO_2 exposure is not well defined. Nevertheless, there are some early studies using NO_2 concentrations >1.5 ppm (2.8 mg/m^3) which resulted in macrophage and neutrophil recruitment into rodent lungs.[80] Similarly, in a recent study in which ferrets were exposed to up to 20 ppm (37.6 mg/m^3) NO_2 for 4 hours, there was a dose-related influx of macrophages and neutrophils over the first 24 hours.[81] In addition, there was a transient increase in eosinophils 48 hours after exposure which coincided with oedema, but without protein leakage into lung lavage fluid. Although these latter findings suggest induction of an inflammatory response that could be related to, or significant in, development of airway hyperreactivity, this unfortunately was not assessed.

4.78 In another study,[82] exposure of guinea pigs to lower levels of NO_2 (4 ppm, 7.5 mg/m^3; 24 hours/day, 1, 3 and 7 days) was shown to induce transient hyperreactivity following histamine aerosol challenge. This was apparent by day 3 but had disappeared by day 7. This effect was inhibited by a thromboxane synthetase inhibitor suggesting that thromboxane might have been an important mediator. In support of these observations, a similar study in which guinea pigs were exposed to either 3 or 9 ppm (5.6 or 17 mg/m^3) NO_2 for 6 hours/day, 6 days/week for 2 weeks, resulted in eosinophil accumulation and epithelial injury which also involved reduced ciliary activity.[83]

4.79 An important difference between these later NO_2 studies and those with ozone is that longer term exposures to NO_2 were required to produce an effect. Shorter exposures of Sprague-Dawley rats to NO_2 (5 ppm, 9.4 mg/m^3; 24 hours) did not induce a change in the response of bronchial rings to acetylcholine or serotonin *ex vivo*,[84] although similar exposure regimes (10 ppm, 18.8 mg/m^3 NO_2; 4–24 hours) resulted in protein transudation into BAL fluid and an increase in BAL, but not mucosal, neutrophil numbers.[85]

4.80 Brief exposures of sheep to very high concentrations of NO_2 (100 or 500 ppm; 188 or 940 mg/m^3 for 15 minutes) caused increased alveolar and capillary leukocytes and an increase in lung resistance, respiratory rate and minute ventilation (ie, an irritant response) although hyperreactivity was not measured.[86] Without detailed time-course studies in which cellular and physiological measurements are made in parallel, it remains unclear whether the hyperreactivity observed after NO_2 exposure involved direct action on nerves and muscle, or whether the cellular inflammation is a pre-requisite of NO_2-induced hyperreactivity, or alternatively if, like ozone, both mechanisms can occur to induce an enhanced hyperreactive response.

Sulphur dioxide

4.81 Sulphur dioxide (SO_2) is known to be a potent bronchoconstrictor and the likely underlying mechanisms have been addressed in detail in a previous report.[23] Thus, SO_2 has been shown in numerous animal studies to interact directly with the sensory nerves in the airway mucosa. These nerves are readily accessible to inhaled chemicals, since the terminals lie within the epithelium, often just below epithelial cell tight junctions. Access might be enhanced following direct damage to epithelial cells. Animal studies showing that SO_2 exposure induces both bronchoconstriction and cough suggest that the two main types of SO_2-sensitive receptors are C-fibres, located in the airway walls and alveoli, and rapidly adapting "irritant" receptors, found in the larynx and large airways. In addition, activation of these receptors will induce mucus secretion from sub-mucosal glands as well as mucosal vasodilation, which will add to airway obstruction.

4.82   Neurogenic inflammation may occur following exposure to SO_2. Activation of sensory nerves could result in the release of neuropeptides including substance P, neurokinins and calcitonin gene-related peptide in the vicinity of airway smooth muscle, mucosa and mucosal glands leading to bronchoconstriction, vasodilation and mucus secretion.

4.83   In support of these suggestions, exposure of isolated perfused guinea pig lungs to a high concentration of SO_2 (250 ppm, 715 mg/m³) induced bronchoconstriction which could be reversed by ruthenium red and partially reversed by lidocaine.[87] It could not be reversed by indomethacin (a cyclooxygenase inhibitor) or diphenhydramine (an H1-receptor antagonist). SO_2 also induced release of calcitonin gene-related peptide but not histamine or thromboxane. Thus, the mechanism of action of SO_2-induced bronchoconstriction was ascribed to activation of sensory nerves (C-fibres) via calcium-dependent release of sensory neuropeptides (as indicated by the inhibitory action of ruthenium red) rather than via inflammatory cell mediators. Pretreatment of such lung preparations with a lower, possibly more relevant, concentration of SO_2 (10 ppm, 28.6 mg/m³ for 30 minutes) did not induce bronchoconstriction but protected against bronchoconstriction induced by 250 ppm (715 mg/m³) SO_2.[88] Inclusion of sodium sulphite in the preparation was also protective and the authors suggest that formation of sulphite during the first exposure may have rendered the airways less susceptible to later exposure.

4.84   It is also possible to induce transient hyperreactivity with SO_2. However, extremely high concentrations of SO_2 have been used. Thus, exposure of dogs to 200 ppm or 400 ppm (572 or 1144 mg/m³) for 2 hours induced airway hyperreactivity to histamine. This was directly associated with sloughing of the epithelium (within 15 minutes) and an increase in bronchoalveolar lavage fluid plasma proteins and PGE_2.[89] An early neutrophil influx occurred between 1 and 4 hours following exposure[90] and by 24 hours there was also an increase in bronchoalveolar lavage macrophage numbers.[89] There appears to have been no change in mast cells or eosinophils at these time intervals.

4.85   Thus, while there is little doubt that in animal models SO_2 can induce bronchoconstriction, hyperreactivity and airway inflammation, high doses have usually been used and where lower dosing regimes have been studied, low doses of SO_2 (10 ppm, 28.6 mg/m³; 30 minutes) did not cause bronchoconstriction and, on the contrary, had a "protective" effect with regard to subsequent exposure to a higher dose of SO_2 (250 ppm, 715 mg/m³).[88] Humans are more sensitive to SO_2 and react at concentrations below 5 ppm (14.3 mg/m³); asthmatic patients are more sensitive than healthy subjects. Exposure of animals to concentrations of SO_2 similar to those relevant in man results in little obvious damage.[23] The consistent observation of epithelial damage/shedding may be significant during combined exposure to SO_2 with other agents (see below).

Acid aerosols

4.86   Published studies concentrate mostly on the effects of sulphuric acid. Ammonium sulphate and ammonium bisulphate generated after partial neutralisation of sulphuric acid by airborne and airway ammonia, are generally less toxic.[91]

4.87   The cellular interaction with acid aerosols is critically dependent on the size of the droplets and hence the site of deposition in the lung. In addition hygroscopic growth may alter the pattern of deposition.[92] In the studies described below, the particles were usually fine (less than 0.6 μm mass median diameter, MMD) or ultrafine (less than 0.1 μm MMD) and would be expected to interact with both conducting and respiratory regions of the respiratory tract. Studies by Pierson et al and by Friedlander[93,94] suggest that atmospheric acid aerosol particles have a MMD of approximately 0.3 μm.

4.88 Both chronic and acute exposure to sulphuric acid have been shown to increase airway sensitivity. Unlike the response induced by ozone, although sensitivity can increase during and immediately after exposure to acid, the response does not appear to persist. Chronic exposure of rabbits to relatively high levels of sulphuric acid aerosol (250 μg/m³) for 1 hour/day, 5 days/week, increased the airway sensitivity to iv acetylcholine after 4 months exposure, which increased further after 8 and 12 months exposure.[95]

4.89 Transient hyperresponsiveness was observed following exposure of guinea pigs to very high (3.2 mg/m³, 24 hour/day) levels of sulphuric acid, which was apparent after 14 days, but not before (3 or 7 days) or after (30 days) this.[96] Interestingly, the same exposure protocol induced transient hyporesponsiveness after 3 days exposure. Acute exposure of guinea pigs to sulphuric acid (200 μg/m³) for just one hour caused an immediate increase in airway sensitivity to intravenous acetylcholine.[97] Acute exposure to more relevant doses (3 hours, 50–500 μg/m³) have since been shown[98] to cause hyperresponsiveness following ex vivo challenge of bronchial tracheal rings to histamine or acetylcholine, which was apparent at exposure >75 μg/m³. These latter observations may have relevance since similar atmospheric concentrations of sulphuric acid (75 μg/m³) have been estimated to occur during peak exposure times (1 hour) and exposure of asthmatic patients to 70 μg/m³ of sulphuric acid has been shown to affect pulmonary function.[99]

4.90 There is little evidence of gross epithelial damage or inflammatory cell influx following sulphuric acid inhalation, particularly at the lower exposure levels.[91] In fact, there appears to be few, if any, long-term cellular effects of sulphuric acid even at very high exposure levels. For example, histopathological examination of airway tissue sections at 4, 8 and 12 months, showed no evidence of epithelial sloughing or inflammatory cell influx, although there was focal hyperplasia, following chronic exposure of rabbits to relatively low levels of sulphuric acid (125 μg/m³, 2 hours/day, for up to a year).[100] Neither was there evidence of an acute effect of SO_2 on epithelial cell integrity or inflammatory cell influx into rabbit bronchial immediately after exposure to a wide range of sulphuric acid concentrations (50–500 μg/m³)[98] or in rabbit bronchoalveolar lavage immediately after exposure to between 50 and 125 μg/m³ sulphuric acid for 3 hours.[101]

4.91 These studies are in agreement with an earlier study in which Cynomolgus monkeys were exposed to relatively high doses of sulphuric acid (0.38–4.79 mg/m³) for 78 weeks and in which guinea pigs were exposed to 0.08 or 0.1 mg/m³ sulphuric acid; there was either no effect (guinea pigs) or focal epithelial changes (thickening of the alveolar and respiratory bronchiolar regions in Cynomolgus monkeys).[102] There are no detailed microscopic or bronchoalveolar lavage studies showing an effect over time of sulphuric acid exposure on inflammatory cell influx into the lung.

4.92 The mechanism by which sulphuric acid causes bronchoconstriction and hyperresponsiveness is unclear. The absence of inflammatory cell influx and any prolonged effects on airway sensitivity suggest that sulphuric acid has a direct impact on airway cells and tissue components that is manifested only during exposure. El-Fawal and Schlesinger suggest that there may be acid-induced alteration of H_3 (dilatory) histamine receptors or other receptors that modulate constrictor responses.[98] Alternatively, they suggest that increased acidity might affect release of other important mediators of bronchoconstriction, such as PGE_2. Activation of resident mast cells may also contribute since histamine release by mast cells isolated from guinea pig lungs immediately after exposure to sulphuric acid is elevated. This effect was dependent on dose and time of exposure (1 and 3.2 but not 0.3 mg/m³, for 2 weeks, but disappeared by 4 weeks).[103]

4.93   Thus, the pulmonary effects of sulphuric acid aerosols depend on particle size, dose and duration of exposure. Variability between experiments likely reflects differences in these parameters and species differences. However, it is clear that acid aerosols do not cause the same degree of epithelial cell toxicity/damage nor inflammatory cell infiltration seen with oxidants. Furthermore, bronchoconstriction and bronchial hypersensitivity are apparent during, but do not persist after, exposure. It is interesting that both acute and chronic exposure will increase airway reactivity as this suggests that either acute or chronic exposure to sulphuric acid could contribute to human asthma. Sulphuric acid inhalation can cause increased or decreased bronchociliary clearance depending on dose, and an increase in secretory cell numbers, depending on the exposure regime;[100,104] it seems possible that a build up of respiratory secretions might protect the airways and account for hyporeactivity observed during some exposure regimes and some of the variation in response over time and between studies.

4.94   There is little information, however, on the metabolic effects of sulphuric acid on pulmonary epithelium. Since sulphuric acid induces a significant decrease in the intracellular pH of rodent lung macrophages[105,106] and a human tracheal epithelial cell line and also compromises many aspects of macrophage function and TNF–α production,[106,107] one could hypothesise that sulphuric acid might precipitate changes in intracellular metabolism of epithelial cells, for example, cytokine synthesis and release, that may be significant in terms of airway reactivity, particularly in concert with other inhaled gases and aeroallergens.

Particles

4.95   Epidemiological studies show a strong relationship between the amount of airborne particles and the incidence of respiratory disease. However, there is much debate concerning exactly how the loading of the particles (ie, concentrations, chemical composition, size etc) corresponds to lung disease. A detailed review of the effects of particles upon health is available.[108]

Deposition of particles in the respiratory tract

4.96   Inhaled antigenic molecules, allergens, are generally contained in or carried on the surface of particles or droplets. Allergens may be soluble in liquid droplets or, perhaps, in the surface film which coats many droplets in the airways. Deposition of allergen carried in this way will be dependent upon the same mechanisms as those controlling deposition of the particles or droplets *per se.*

4.97   A detailed account of the factors controlling deposition of particles in the respiratory tract was included in the report of the Advisory Group on the Medical Aspects of Air Pollution Episodes dealing with Sulphur Dioxide, Acid Aerosols and Particulates.[92] Here only a brief summary is provided.

4.98   A very large variety of particles are found in ambient air. The range of particle types reflects the range of sources including emissions from motor vehicles, wind blown dust, particles formed by reactions taking place in the gas phase and organic sources such as plants. Pollens will be discussed in more detail below. The abundance of particles in the air may be described in a number of ways. Particle counts are perhaps the simplest: in relatively unpolluted urban air some 10,000 particles might be found in each cm^3 of air. This concentration rises under polluted conditions and under conditions of intense pollution over 150,000 particles might be found in each cm^3.

4.99   Ambient air contains particles of varying size. The distribution of particle sizes may be described in a variety of ways. A special terminology has been developed to describe such approaches: details may be found in an account by Hinds.[109] Some of the standard terms used are shown in Figure 4.5. The importance of large particles as far as the distribution of mass in the aerosol is clear from Figure 4.5: 50% of the mass of the aerosol is contained in particles of less than 30 μm diameter (MMD = 30 μm). In terms of numbers of particles on the other hand, half the total number of particles are of less than 0.9 μm diameter (Count Median Diameter=9.0 μm).

Figure 4.5 **Arithmetic plot of size data for an aerosol of CMD=9.0 *µ*m; GSD (σg)=1.89 *µ*m, showing various average diameters.**

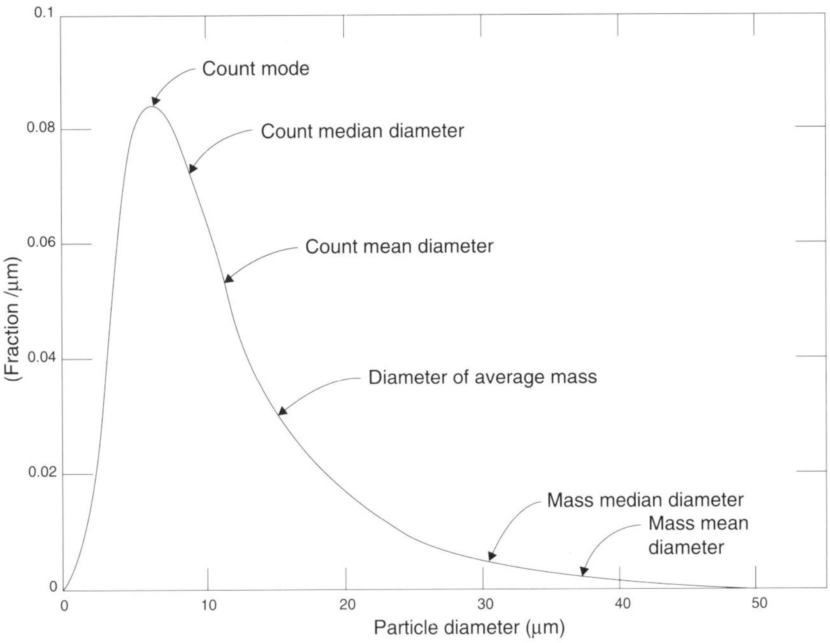

4.100   Particles inspired via the mouth or the nose are deposited in the respiratory tract as a result of four main mechanisms:

Sedimentation

4.101   Particles suspended in the air fall under the influence of gravity and settle on surfaces of the airways. The rate of settling is dependent upon both particle density and the square of the particle radius.

Inertia

4.102   Particles possess inertia and tend to continue in their direction of movement when the air stream deviates as a result of bending and branching of the airways. Inertial deposition is dependent upon particle density and the square of particle radius.

Diffusion

4.103   Small particles are significantly affected by impact of gas molecules undergoing Brownian motion. Diffusion of particles results and is dependent on particle radius but is independent of particle density. Very small particles diffuse rapidly, though of course, very much more slowly than gas molecules.

Interception

4.104   Fibres travelling in an airstream impact at bends and branch points in the airways because they are unable to follow the change in the direction of air flow, travel tangentially and encounter airway walls.

4.105   These mechanisms lead to differential deposition of particles in the airways. Experimental studies have generally confirmed the results of theoretical modelling of patterns of deposition. Particles of greater than 5 *µ*m diameter are very effectively filtered by the respiratory system mainly by the upper respiratory tract. This is clearly shown in Figure 4.6.

Deposition at the alveolar level peaks at about 3 *µ*m during mouth breathing and at about 2 *µ*m during nasal breathing. This is shown in Figure 4.6.

Figure 4.6 **Experimental data for deposition in the alveolar region. Deposition is expressed as fraction of mouthpiece inhalation versus aerodynamic diameter (geometric diameter below 0.5 μm).**

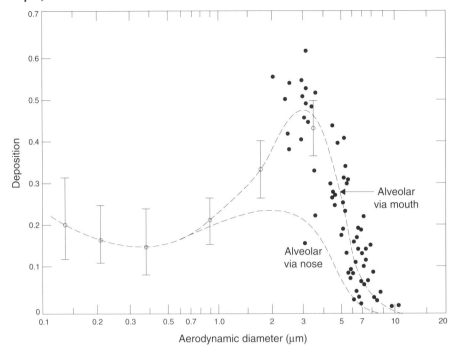

4.106 Devices designed to sample particles likely to reach the alveoli are designed to accept particles of size less than about 5 μm diameter. A number of such devices have been developed. Figure 4.7 shows the acceptance curves for precollectors designed by the American Conference of Governmental Industrial Hygienists (ACGIH) 1968 and the British Medical Research Council (BMRC) 1952.

Figure 4.7 **Comparison of experimental measurements of alveolar deposition, Task Group model, and precollector penetration.**

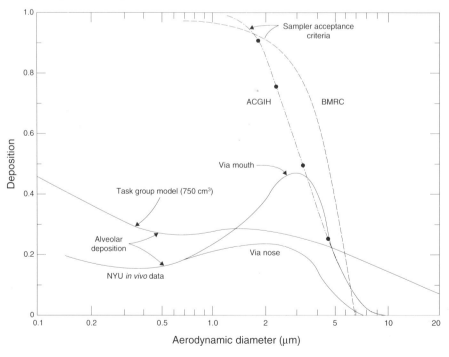

It will be seen that the curves relating to the alveoli give estimates of the percentage of particles of a certain size reaching and being deposited in the alveoli, the curves for the collection devices give the percentage of particles passing the size selective collector.

4.107   For some purposes, an estimate of the percentage of particles deposited in the alveolar region and the conducting airways, excluding the nasopharyngeal region, is required. Such particles are referred to as "Thoracic Particles" and are monitored using a device which collects particles of less than 10 μm diameter: this sample is referred to as PM_{10}. The performance of the device is more formally defined in terms of a 50% acceptance of particles of 10 μm diameter. Like the acceptance curves shown in Figure 4.7 the acceptance curve for a PM_{10} sampling head is steep: almost 100% of particles of less than 9 μm diameter are collected and very few of diameter >11 μm are collected.

Allergens in ambient air

4.108   A wide range of pollens and spores are found in ambient air both indoors and out. These range in size from some tens of microns diameter for the grass pollens to just a few microns diameter for fungal spores. Large pollen grains are deposited in the nose where soluble antigens are released and may set up inflammatory reactions. Smaller pollens and spores may be inhaled more deeply into the lungs. Responses to the deposition of pollens and spores can vary greatly from individual to individual. In some, rhinitis is the predominant effect, in others, asthma predominates. Pollens and spores also vary greatly in their capacity to induce allergenic responses.

4.109   Pollen grains may be broken down into smaller allergenic fragments in the air. This process is enhanced during thunderstorms, and reports have appeared in the literature of outbreaks of asthma following such storms. It has also been suggested that fragments of pollen grains, and presumably spores, may be carried deep into the lung on the surface of other particles found in the ambient urban aerosol. This suggestion has arisen as a result of the observations that sensitisation to cedar pollen was more likely in Japanese subjects living near to busy roads than in those living further from roads but equally close to trees (see Chapter 9). Animal studies have shown that particulate material collected from diesel exhaust enhances the allergic response to both intraperitoneally and intranasally administered ovalbumin.

Animal studies on the effects of particulate material

4.110   The site of deposition of inhaled particles in animal models depends on the aerodynamic diameter of a particle. Thus, particles of diameter between 0.05 and 10 μm (ie, corresponding to that thought to be significant to human disease), which increase during episodes of air pollution, would be expected to be deposited throughout the conducting and respiratory airways.

4.111   A recent study demonstrated a rapid increase (1.5-fold) in airway macrophages (within 20 minutes) following inhalation of 6 μm polystyrene particles. Macrophage numbers continued to increase (3-fold) for up to 40 minutes, but had started to decline (2 times control) 27 hours following exposure.[110] It seems possible that rapid congregation of macrophages and release of mediators in the airways may be significant during episodes of air pollution where both particles and other pollutants may be interacting and cause an asthmatic response. In addition, very fine particles are apparently able to penetrate the epithelial wall and may have alternative, as yet unknown, mechanisms of action.[111,112]

4.112   Although most experimental animals studies demonstrate a reasonably clear-cut relationship between particle size and site of action of inhaled particles, the exact response also depends on the subsequent chemical and cellular interactions. Thus, during episodes of air pollution, a critical factor is likely to be not only the particle load but also their composition and the nature of the chemicals "delivered" in parallel with the particles. This is discussed later in this chapter.

Air pollutants and allergy

4.113   There is an interesting body of evidence to suggest that air pollutants enhance allergic sensitisation. Some early studies by Matsumura examined the effects of ozone, NO_2 and SO_2 on induction of respiratory allergy in guinea pigs.[113,114] Repeated exposure (5–6 times, 30 minute exposure) to each of the three gases enhanced sensitisation to inhaled antigen (egg albumin mixed with bovine serum albumin) which was carried out by nebuliser 30–50 minutes after gas exposure. This was apparent at the higher (ozone 5 ppm, 10 mg/m³; NO₂ 70 ppm, 132 mg/m³; SO₂ 330 ppm, 944 mg/m³), but not at lower, dose levels. The presence of antibody to egg albumin in a high proportion of anaphylactic test animals suggested that exposure to the pollutants had promoted airway sensitisation.[113] Subsequently, exposure to 8 ppm (16 mg/m³) ozone for 30 minutes was shown to significantly retard clearance of inhaled egg albumin from the lung (via bronchociliary action) with a concomitant rise of antigen in the blood, possibly due to disruption of airway epithelium, in effect increasing the antigen load and possibly explaining the increased antigenic response seen in the previous study.[114] Similar mechanisms may account for the effects observed following exposure of mice to sulphuric acid after prior exposure to ozone and allergen. The enhanced sensitivity to allergen that occurred following ozone exposure was found to be further enhanced by additional exposure to sulphuric acid.[115]

4.114   Although in the studies by Matsumura, SO₂ was not the most potent of the gases tested, Riedel *et al* have since shown that much lower doses of SO₂ (5 ppm, 14.3 mg/m³) enhanced allergic sensitisation when guinea pigs were exposed to SO₂ for 5 days, 8 hours/day.[116] The animals were intermittently exposed to ovalbumin aerosol (or air) during this period. When the animals underwent bronchial provocation with nebulised ovalbumin 7 days after exposure, SO₂-exposed animals were found to be sensitised, whereas control animals, and animals treated with methylprednisolone, indomethacin and inhaled nedocromil sodium during the exposure period, were not. Histological examination showed a marked inflammatory response throughout the airways of animals exposed to SO₂: the predominant inflammatory cells were polymorphonuclear neutrophils particularly in bronchioles. There was also severe epithelial damage in these airways, which was less obvious in the main bronchi and trachea. In contrast, alveoli were filled with macrophages and monocytes. Treatment with anti-inflammatory drugs prevented the peripheral inflammation and tissue damage, although the milder, upper airway damage was not altered. Antibody titres to ovalbumin increased only in the animals exposed to SO₂.

4.115   This study suggests that SO₂-induced epithelial damage and inflammation significantly enhance the allergenicity of aeroallergens, possibly allowing greater access of antigen to immune effector cells. However, the histological study was carried out on day 5 of the exposure; there is no information on airway cell integrity or inflammation on the day of ovalbumin challenge when damaged epithelium may still have been present and other inflammatory cells might have, at least in part, accounted for some of the observed increased response.

4.116   Recent studies of the interaction between allergen and ozone in the induction of hyperreactivity support the concept of enhanced early phase response following subsequent challenge. Actively sensitised guinea pigs (with ovalbumin) were exposed to ozone (30 minutes or 1 hour, 3–5 ppm; 6–10 mg/m³) then challenged with histamine,[117] ovalbumin,[117] or 5-hydroxytryptamine.[118] In both studies, bronchial responsiveness was increased compared with controls. It was suggested that a selective effect of ozone on resident, ovalbumin-induced inflammatory cells potentiates the response to subsequent challenge. However, in similar studies in *Ascaris suum*-sensitive mongrel dogs, Kleeberger and colleagues[119] noted that the immediate response to antigen challenge was attenuated or unaffected by a brief ozone exposure (1 ppm, 2 mg/m³), depending on dose of antigen used. Furthermore, they found that pre-exposure to 1 ppm (2 mg/m³) ozone for 5 minutes completely blocked the late phase response to antigen delivered 3 hours after ozone.[120] This was despite the presence of increased numbers of mast cells and neutrophils in the lung at this time as a result of the prior exposure to ozone.

4.117   The studies discussed above indicate that exposure to each of the three major air pollutants gases: ozone, NO₂ and SO₂ can affect both the likelihood of sensitisation to inhaled allergen and the response seen in sensitised animals on allergen challenge. The majority of studies have involved exposure to concentrations of pollutants significantly, in some cases very significantly, greater than levels encountered in ambient air in the UK. The effects of exposure to pollutant gases on early and late responses have been shown, in some models, to differ. The theory that epithelial damage, by pollutants, enhances the likelihood of sensitisation is supported. It is clear, also, that some pollutants can affect the effector arm of the allergic response. Extrapolation from these findings to likely effects in man is difficult as significant species differences have been demonstrated.

Adjuvant effect of diesel exhaust particles

4.118   Diesel exhaust particles (DEP) are small, easily respirable and studies in experimental animals show that, once deposited in the lung, they are translocated to lung-associated lymph nodes.[121] Immune cells within these lymph nodes have been shown to proliferate and produce antibody following immunisation with antigens such as sheep red blood cell preparations.[122] Thus, DEP deposition and lymph node cellularity increased after exposing rats and mice to 7000, 3000 and 350 μg DEP/m^3, 7 hours/day, 5 days/week for 6, 12, 18 and 24 months. However, there was only a slight increase in the immune IgM response in this study, although IgE was not measured.

4.119   In this respect, DEP has since been shown to act as an adjuvant for the production of IgE.[123] The nasal passages of mice were inoculated with 0, 1, 5 or 25 μg DEP in 25 μl borate buffered saline containing 0, 0.025, 0.25 or 2.5 μg ovalbumin and boosted 4, 7, 10 and 13 weeks after initial exposure. These doses and route of exposure could be considered realistic as far as human exposure is concerned. DEP significantly enhanced the capacity of ovalbumin to trigger an IgE antibody response.

4.120   This type of study has been extended to humans,[124] and it is worth mentioning that nasal challenge with DEP (0.3 mg) led to an increase in the number of IgE-producing cells as well as IgE levels (1.4–10.2-fold) in nasal lavages 4 days, but not 7 or 10 days, after nasal challenge. IgG$_4$ was also increased in nasal lavage, but this was seen only on the 7th day after challenge. Other Ig isotypes remained normal in nasal lavage and all Ig isotypes, including IgE, remained at constant levels in serum from exposed subjects. Exposures to higher or lower levels of DEP (1 μg or 0.15 μg) had no effect. These data suggest that DEP can stimulate local IgE production by nasal immune cells and hence may be significant in respiratory allergic disease. It is clear from this study that exposure level (equivalent to breathing outdoor air in Los Angeles for 24 hours) is critical and that the duration of the response is finite.

4.121   Little is known about the mechanisms involved in this process, although DEP has been shown to have direct effects on T and B cells, possibly stimulating numbers of IgE producing B cells and specifically inducing B cells to synthesise IgE. DEP may also stimulate epithelial cells to synthesise and release stimulatory factors (eg, IL–4) in a similar manner.[124] There was no evidence, in this study, for increased epithelial permeability (no increase in albumin levels) and enhanced access of DEP to underlying tissue/cells, suggesting "non-toxic", cell-specific interactions between DEP and nasal mucosa. This seems possible, since DEP is a complex mixture of compounds with a carbonaceous core and many adsorbed organic components.

4.122   Diesel particles are small (<0.5 μm) and therefore have a high surface area to mass ratio. Thus, small amounts of material would provide significant surface areas of potentially highly reactive material to resident cells such as T and B cells, mast cells, macrophages and epithelial cells.

Combined exposure protocols and additive, synergistic, antagonistic and potentiative effects

4.123   Exposure to a combination of oxidant gases and/or acid aerosols with particles may prove to be a critical factor in the relative pathophysiological effects of pollutants. For example, Schlesinger and colleagues have investigated how combined exposure to sulphuric acid and ozone affects a variety of parameters in the lungs of rabbits.[100] Chronic exposure to sulphuric acid (125 μg/m^3) and ozone (0.1 ppm, 0.2 mg/m^3) for 2 hours/day for up to 1 year, had both synergistic and antagonisic effects on tracheobronchial mucociliary clearance and epithelial secretory cell number, depending on the time of analysis during exposure (synergistic at 4 months, antagonistic by 8 months).

4.124   In a more complex study where rabbits were exposed to 4 concentrations of each of the following: sulphuric acid 1, 50, 75 and 125 $\mu g/m^3$; ozone 0, 0.1, 0.3 and 0.6 ppm (0, 0.2, 0.6 and 1.2 mg/m^3) for 3 hours in every combination, many parameters were examined, including macrophage phagocytosis, superoxide production, prostaglandin and TNF production and inflammatory cell profile.[101] The combined exposure protocol either had no effect or there was significant antagonism between sulphuric acid and ozone (with the exception of TNF release which was elevated at the highest doses of acid and ozone) which was not related to exposure concentrations or combinations of pollutants.

4.125   A similar antagonistic response was observed following exposure of rats to nitric acid vapour (1.0 mg/m^3) and ozone (0.6 ppm, 1.2 mg/m^3) for 4 hours or to nitric acid (0.25 mg/m^3) and ozone (0.15 ppm, 0.3 mg/m^3) for 4 hours/day, for 4 days.[125] The effects of combined exposure on bronchoalveolar lavage fluid protein levels, inflammatory cells, leukotriene production and macrophage respiratory burst was either the same or less than that seen after exposure to each pollutant alone. No synergism was demonstrated. These studies did not address bronchoconstriction, the mechanism of which can be independent of inflammatory cell action and might, therefore, respond differently in combined exposure protocols.

4.126   It is of interest that in studies on the effects of acid aerosols in animal models, comparatively high doses are often required to bring about a response, whereas studies in asthmatics suggest that relatively low concentrations of sulphuric acid (70 $\mu g/m^3$) will elicit adverse pulmonary effects. As mentioned earlier, variability between experimental animal studies may be due to the size or format of the delivered aerosols. In this respect, Amdur and colleagues have carried out extensive studies on the possible additive or synergistic effects of ultrafine particles on the action of inhaled sulphuric acid. This is relevant during coal combustion or smelting, when metal oxides react with SO_2 to generate acid-coated, ultrafine particles. Exposure of guinea pigs to 20 $\mu g/m^3$ of ZnO particles coated with a surface layer of sulphuric acid, 3 hours/day, over 5 days, resulted in reduced lung function with increased bronchoalveolar lavage protein and neutrophil numbers.[97] Significantly, only a single 1 hour exposure to the same dose of sulphuric acid-coated ZnO induced bronchial hyperreactivity, whereas a far more concentrated (200 $\mu g/m^3$) suspension of sulphuric acid aerosol of equivalent droplet size was required to induce the same degree of hyperresponsiveness.

4.127   Similarly, when airway hyperresponsiveness to acetylcholine was examined 90 minutes after 1 hour exposure to sulphuric acid-coated metal oxide particles (30 $\mu g/m^3$ sulphuric acid, 2.3 mg/m^3 ZnO), it was significantly increased compared to that of animals exposed to 2.8 mg/m^3 ZnO or 1.0 ppm (2.86 mg/m^3) SO_2, which did not differ significantly from control levels.[126] As might be expected, a 10-fold greater amount of sulphuric acid aerosol was required to induce the same change in airway hyperresponsiveness. The effect of prior exposure to sulphuric acid-coated particles on the effect of either ozone or repeated acid-coated particle exposure on guinea pig lung function was also studied. Thus, exposure to 24 $\mu g/m^3$ sulphuric acid layered on to 1 mg/m^3 ZnO for 3 hours/day over 5 days was found to sensitise the animals so that further exposure (day 9) to 0.15 ppm (0.3 mg/m^3) ozone or a repeat exposure to acid-coated particles for 1 hour caused a significant decline in diffusing capacity and vital capacity that was not seen with either agent alone. Thus, inhalation of acid-coated particles potentiated the effect of ozone.[127]

4.128   These studies illustrate how the format of inhaled pollutants can be crucial to their effects. In the case of acid-coated particles, it appears that the ultrafine particles might act as vectors, delivering acid to the lung in a way that maximises its effects by enhancing surface area to mass ratio and perhaps delaying clearance and detoxification of the acid. Retention in the lung may be particularly important where subsequent exposure to the same or an alternative pollutant takes place and thus may potentiate an additive or synergistic response. It would be interesting to examine inhalation of acid-coated particles in concert with other pollutants, such as NO_2, or with allergens to examine allergic sensitisation. A variety of exposure protocols might also establish whether sequence of exposure is critical to the magnitude and type of effect.

4.129   The studies considered in this chapter support the suggestion that exposure to pollutants, both gases and particles, can increase the likelihood of sensitisation to allergen. The response of sensitised animals to antigen challenge can also be increased by exposure to pollutants.

4.130   Of the responses studied, the enhanced sensitisation produced by exposure to diesel exhaust particles and the effects of very small metal-containing, and acid-coated, particles are, perhaps, the most interesting. These responses have been demonstrated at low levels of exposure and may well be relevant to man at ambient levels of exposure.

References

1.   Holgate ST, Beasley R, Twentyman OP. The pathogenesis and significance of bronchial hyperresponsiveness in airways disease. Clin Sci 1987; 73:561–572.

2.   Djukanovic R, Roche WR, Wilson JW, Beasley CRW, Twentyman OP, Howarth PH, Holgate ST. Mucosal inflammation in asthma. Am Rev Respir Dis 1990; 142:434–457.

3.   Laitinen LA, Laitinen A, Haahtela T. Airway mucosal inflammation even in patients with newly diagnosed asthma. Am Rev Respir Dis 1993; 147:697–704.

4.   Corrigan CJ. Allergy of the respiratory tract. Curr Opin Immunol 1992; 4:798–804.

5.   Busse WW, Calhoun WF, Sedgwick JD. Mechanism of airway inflammation in asthma. Am Rev Respir Dis 1993; 147:520–524.

6.   Barnes PJ, Chung KF, Page CP. Inflammatory mediators and asthma. Pharmacol Rev 1988; 40:49–84.

7.   Bradding P, Roberts JA, Britten KM, Montefort S, Djukanovic R, Müller R, Hüsser CH, Howarth PH, Holgate ST. Interleukin–4, –5 and –6 and tumor necrosis factor–α in normal and asthmatic airways: evidence for the human mast cell as a source of these cytokines. Am J Respir Cell Mol Biol 1994; 10:471–480.

8.   Corrigan CJ, Kay AB. T cells and eosinophils in the pathogenesis of asthma. Immunol Today 1992; 13:501–506.

9.   Rochester CL, Rankin JA. Is asthma T-cell mediated? Am Rev Respir Dis 1991; 144:1005–1007.

10.   Holgate ST, Djukanovic R, Howarth PH, Montefort S, Roche W. The T cell and the airway's fibrotic response in asthma. Chest 1993; 103:125S–129S.

11.   Jeffery PK, Wardlaw AJ, Nelson FC, Collins JV, Kay AB. Bronchial biopsies in asthma: an ultrasonic quantitative study and correlation with hyperactivity. Am Rev Respir Dis 1989; 140:1745–1753.

12.   Azzawi M, Bradley B, Jeffrey PK, Frew AJ, Wardlaw AJ, Knowles G, Assoufi B, Collins JV, Durham S, Kay AB. Identification of activated T lymphocytes and eosinophils in bronchial biopsies in stable asthmatics. Am Rev Respir Dis 1990; 142:1407–1413.

13.   Corrigan CJ, Hartnell A, Kay AB. T lymphocyte activation in acute severe asthma. Lancet 1988; i:1129–1131.

14.   Chatelain R, Varkila K, Coffman RL. IL–4 induces a Th2 response in *Leishmania major*-infected mice. J Immunol 1992; 148:1182–1187.

15.   MacDermot J, Fuller RW. Macrophages. In: Barnes P, Rodger I, Thomson N, editors. Asthma: Basic Mechanisms and Clinical Management. London: Academic Press, 1988; 87–113.

16. Pueringer RJ, Hunninghake GW. Inflammation and airway reactivity in asthma. Am J Med 1992; 92:32S–38S.

17. Sur S, Crotty TB, Kephart GM, Hyma BA, Colby TV, Reed CE, Hunt LW, Gleich GJ. Sudden-onset fatal asthma: a distinct entity with few eosinophils and relatively more neutrophils in the airway submucosa? Am Rev Respir Dis 1993; 148:713–719.

18. Davies RJ, Devalia JL. Air pollution and airway epithelial cells. Agents Actions Suppl.1993; 43: 87–96.

19. Cromwell O, Hamid Q, Corrigan CJ, Barkans J, Meng Q, Collins PD, Kay AB. Expression and generation of interleukin–8, interleukin–6 and granulocyte-macrophage colony-stimulating factor by bronchial epithelial cells and enhancement by IL–1β and tumour necrosis factor–α. Immunology 1992; 77:330–337.

20. Vanhoutte PM. Epithelium derived relaxing factors(s) and bronchial reactivity. Am Rev Respir Dis 1988; 138(Suppl):S24–S30.

21. Barnes PJ, Baraniuk JN, Belvisi MG. Neuropeptides in the respiratory tract. Part I. Am Rev Respir Dis 1991; 144:1187–1198.

22. Devalia JL, Wang JH, Sapsford RJ, Davies RJ. Expression of RANTES in human bronchial epithelial cells down regulated by beclomethasone dipropionate (BDP). Presented at the British Society of Allergy and Clinical Immunology Annual Conference, 1994.

23. Department of Health. Advisory Group on the Medical Aspects of Air Pollution Episodes. Second Report: Sulphur Dioxide, Acid Aerosols and Particulates. London: HMSO, 1992.

24. Barnes PJ, Baraniuk JN, Belvisi MG. Neuropeptides in the respiratory tract. Part II. Am Rev Respir Dis 1991; 144:1391–1399.

25. Cross CE, van der Vliet A, O'Neill CA, Louie S, Halliwell B. Oxidants, antioxidants and respiratory tract lining fluids. Environ Health Perspect 1994; 102(Suppl 10):185–191.

26. Department of Health. Advisory Group on the Medical Aspects of Air Pollution Episodes. First Report: Ozone. London: HMSO, 1991.

27. Hatch GE. Comparative biochemistry of the airway lining fluid. In: Parent ED, editor. Treatise on Pulmonary Toxicology. Volume 1. Comparative Biology of the Normal Lung. Boca Raton: CRC Press, 1992; 617–632.

28. Housley D, Mudway I, Kelly F, Eccles R, Richards R. Depletion of urate in human nasal lavage following in vitro ozone exposure. Respir Med 1994; 88:816A.

29. Mudway I, Housley D, Eccles R, Richards RJ, Datta AK, Tetley TD, Kelly FJ. Differential depletion of respiratory tract antioxidants in response to ozone challenge. Respir Med 1994; 88:808A.

30. Smith LJ, Houston M, Anderson J. Increased levels of glutathione in bronchoalveolar lavage fluid from patients with asthma. Am Rev Respir Dis 1993; 147:1461–1464.

31. Dunnill MS. Pulmonary pathology. 2nd edition. London: Churchill Livingstone, 1987.

32. Hogg JC. The pathology of asthma. Chest 1985; 5:152S–153S.

33. Heard BE, Hussain S. Hyperplasia of bronchial muscle in asthma. J Pathol 1973; 110:319–331.

34. Moreno RH, Hogg JC, Pare PD. Mechanisms of airway narrowing in asthma. Am Rev Respir Dis 1986; 133:1171–1180.

35. James AL, Pare PD, Hogg JC. The mechanics of airway narrowing in asthma. Am Rev Respir Dis 1989; 139:242–246.

36. Wiggs BR, Moreno RH, Hogg JC, Hilliam C, Pare PD. A model of the mechanics of airway narrowing. J Appl Physiol 1990; 69:849–860.

37. Lambert RK. Role of bronchial basement membrane in airway collapse. J Appl Physiol 1991; 71:666–673.

38. Laitinen LA, Heino M, Laitinen A, Kava T, Haahtela T. Damage of the airway epithelium and bronchial reactivity in patients with asthma. Am Rev Respir Dis 1985; 131:599–606.

39. Beasley R, Roche WA, Roberts JA, Holgate S. Cellular events in the bronchi in mild asthma and after bronchial provocation. Am Rev Respir Dis 1989; 139:806–817.

40. Dunnill MS. The pathology of asthma, with special reference to changes in the bronchial mucosa. J Clin Pathol 1960; 13:27–33.

41. Hirshmann CA, Downes CA. Experimental asthma in animals. In: Weiss EB, Segal MS, Stein M, editors. Bronchial Asthma. Boston, Toronto: Little, Brown & Company, 1985; 280–299.

42. Karol MH, Griffiths-Johnson DA, Skoner DP. Chemically induced pulmonary hypersensitivity, airway hyperreactivity and asthma. In: Gardner DE, editor. Toxicology of the lung. New York: Raven Press Ltd, 1993; 417–434.

43. Karol MH. Animal models of occupational asthma. Eur Respir J 1994; 7:555–568.

44. Abraham WM, Baugh LE. Animal models of asthma. In: Busse WB, Holgate ST, editors. Asthma and rhinitis. London, Boston: Blackwells, 1995; 961–977.

45. Pretolani M, Vargaftig BB. Commentary: From lung hypersensitivity to bronchial hyperreactivity: what can we learn from animal models? Biochem Pharmacol 1993; 45:791–800.

46. Wanner A, Abraham WM, Douglas JS, Drazen JM, Richerson HB, Sri Ram J. Models of airway hyperresponsiveness. Am Rev Respir Dis 1990; 141:253–257.

47. Biochot E, Lagente V, Carre C, Waltmann P, Mencia-Huerta JM, Braquet P. Bronchial hyperresponsiveness and cellular infiltration in lung from aerosol-sensitised and antigen-exposed guinea pigs. Clin Exp Allergy 1991; 21:67–76.

48. Huston PA, Church MK, Clay TP, Miller P, Holgate ST. Early and late-phase bronchoconstriction after allergen challenge of nonanesthetized guinea pigs. I. The association of disordered airway physiology to leukocyte infiltration. Am Rev Respir Dis 1988; 137:548–557.

49. Haczku A, Moqbel R, Sun J, Jacobson M, Kay AB, Barnes PJ, Chung KF. Effect of a single ovalbumin aerosol exposure on airway responsiveness, inflammatory cell influx into the airways and activation of T cells in the bronchial mucosa of sensitized brown-Norway rats [Abstract]. Am J Respir Crit Care Med 1994; 149(4 Part 2):A759.

50. Smith HR, Ray BS, Irvin CG. A murine model for the late phase pulmonary response, airways hyperresponsiveness and airways inflammation following antigen inhalation [Abstract]. Am Rev Respir Dis 1991; 143:A433.

51. Shampain MP, Larsen GL, Henson PM. An animal model of late pulmonary responses to *Alternaria* challenge. Am Rev Respir Dis 1982; 126:493–498.

52. Metzger WJ. Late-phase asthma in an allergic rabbit model. In: Dorsch W, editor. Late-phase allergic reactions. Boca Raton, Florida: CRC Press, 1990; 347–362.

53. Marsh WR, Irvin CG, Murphy KR, Behrens BL, Larsen GL. Increases in airway reactivity to histamine and inflammatory cells in bronchoalveolar lavage after the late asthmatic response in an animal model. Am Rev Respir Dis 1985; 131:875–879.

54. Chung KF, Becker AB, Lazarus SC, Frick OL, Nadel JA, Gold WM. Antigen induced airway hyperresponsiveness and pulmonary inflammation in allergic dogs. J Appl Physiol 1985; 58:1347–1353.

55. Abraham WM, Delehunt JC, Marchette B. Characterisation of a late phase pulmonary response after antigen challenge in allergic sheep. Am Rev Respir Dis 1983; 128:839–844.

56. Abraham WM, Sielczak MW, Wanner A, Perruchoud P, Blinder L, Stevenson JS, Ahmed A, Yerger LD. Cellular markers of inflammation in the airways of allergic sheep with and without allergen-induced late responses. Am Rev Respir Dis 1988; 138:1565–1571.

57. Gundel RH, Wegner CD, Letts LG. Antigen-induced acute and late-phase responses in primates. Am Rev Respir Dis 1992; 146:369–373.

58. Hamel R, McFarlane CS, Ford-Hutchinson AW. Late pulmonary responses induced by *Ascaris* allergen in conscious squirrel monkeys. J Appl Physiol 1986; 61:2081–2087.

59. Holtzman MJ, Fabbri LM, O'Byrne PM, Gold BD, Aizawa H, Walters EH, Alpert SE, Nadel JA. Importance of airway inflammation for hyperresponsiveness induced by ozone. Am Rev Respir Dis 1983; 127:686–690.

60. Fabbri LM, Aizawa H, Alpert SE, Walters EH, O'Byrne PM, Gold BD, Nadel JA, Holtzman MJ. Airway hyperresponsiveness and changes in cell counts in bronchoalveolar lavage after ozone exposure in dogs. Am Rev Respir Dis 1984; 129:288–291.

61. O'Byrne PM, Walters EH, Aizawa H, Fabbri LM, Holtzman MJ, Nadel JA. Indomethacin inhibits the airway hyperresponsiveness but not the neutrophil influx induced by ozone in dogs. Am Rev Respir Dis 1984; 130:220–224.

62. O'Byrne PM, Walters EH, Gold BD, Aizawa H, Fabbri LM, Alpert SE, Nadel JA, Holtzman MJ. Neutrophil depletion inhibits airway hyperresponsiveness induced by ozone exposure. Am Rev Respir Dis 1984; 130:214–219.

63. Matsui S, Jones GL, Woolley MJ, Lane CG, Gontovnick LS, O'Byrne PM. The effects of antioxidants on ozone-induced airway hyperresponsiveness in dogs. Am Rev Respir Dis 1991; 144:1287–1290.

64. Stevens WH, Adelroth E, Wattie J, Woolley MJ, Ellis R, Dahlback M, O'Byrne PM. The effect of inhaled budesonide on ozone-induced airway hyperresponsivess and bronchoalveolar lavage cells in dogs [Abstract]. Am J Respir Crit Care Med 1994; 149(4 Part 2):A772.

65. Murlas CG, Roum JH. Sequence of pathologic changes in the airway mucosa of guinea pigs during ozone-induced bronchial hyperreactivity. Am Rev Respir Dis 1985; 131:314–320.

66. Evans TW, Brokaw JJ, Chung KF, Nadel JA, McDonald DM. Ozone-induced bronchial hyperresponsiveness in the rat is not accompanied by neutrophil influx or increased vascular permeability in the trachea. Am Rev Respir Dis 1988; 138:140–144.

67. Yeadon M, Wilkinson D, Darley-Usmar V, O'Leary VJ, Payne AN. Mechanisms contributing to ozone-induced bronchial hyperreactivity in guinea-pigs. Pulmon Pharmacol 1992; 5:39–50.

68. Murlas CG, Lang Z, Williams GJ, Chodimella V. Aerosolised neutral endopeptidase reverses ozone-induced airway hyperreactivity to substance P. J Appl Physiol 1992; 72:1133–1141.

69. Okubo T, Nishiyama H, Kaneko N, Yamakawa H, Ikeda H. Ozone induced increase of epithelial permeability of guinea pig trachea, in relation to the airway hyperresponsiveness and increased tracheal capillary permeability [Abstract]. Am J Respir Crit Care Med 1994; 149(4 Part 2):A763.

70. Young C, Bhalla DK. Time course of permeability changes and PMN flux in rat trachea following O_3 exposure. Fundam Appl Toxicol 1991; 18:175–180.

71. Bhalla DK, Young C. Effects of acute exposure to O_3 on rats: sequence of epithelial and inflammatory changes in the distal airways. Inhalation Toxicol 1992; 4:17–31.

72. Bhalla DK, Daniels DS, Luu NT. Attenuation of ozone-induced airway permeability in rats by pretreatment with cyclophosphamide, FPL 55712, and indomethacin. Am J Respir Cell Mol Biol 1992; 7:73–80.

73. Pino MV, Levin JR, Stovall MY, Hyde DM. Pulmonary inflammation and epithelial injury in response to acute ozone exposure in the rat. Toxicol Appl Pharmacol 1992; 112:64–72.

74. Pino MV, Stovall MY, Levin JR, Devlin JR, Koren HS, Hyde DM. Acute ozone-induced lung injury in neutrophil-depleted rats. Toxicol Appl Pharmacol 1992; 114:268–276.

75. Biagini RE, Moorman WJ, Lewis TR, Bernstein IL. Ozone enhancement of platinum asthma in a primate model. Am Rev Respir Dis 1986; 134:719–725.

76. Sielczak MW, Denas SM, Abraham WM. Airway cell changes in tracheal lavage of sheep after ozone exposure. J Toxicol Environ Health 1983; 11:545–553.

77. Kleeberger SR, Seiden JE, Levitt RC, Zhang L-Y. Mast cells modulate acute ozone-induced inflammation of the murine lung. Am Rev Respir Dis 1993; 148:1284–1291.

78. Tan WC, Bethel RA. The effect of platelet activating factor antagonist on ozone-induced airway inflammation and bronchial hyperresponsiveness in guinea pigs. Am Rev Respir Dis 1992; 146:916–922.

79. Tepper JS, Costa DL, Fitzgerald S, Doerfler DL, Bromberg PA. Role of tachykinins in ozone-induced acute lung injury in guinea pigs. J Appl Physiol 1993; 75:1404–1411.

80. Department of Health. Advisory Group on the Medical Aspects of Air Pollution Episodes. Third Report: Oxides of Nitrogen. Chapter 4. Biochemical and Cellular Effects of Oxides of Nitrogen. London: HMSO, 1993; 29–49.

81. Rasmussen RE. Effects of acute NO$_2$ exposure in the weanling ferret lung. Inhalation Toxicol 1992; 4:373–382.

82. Kobayashi T, Shinozaki Y. Induction of transient airway hyperresponsiveness by exposure to 4 ppm nitrogen dioxide in guinea pigs. J Toxicol Environ Health 1992; 37:451–461.

83. Ohashi Y, Kakai Y, Suguira Y, Ohno Y, Okamoto H. Nitrogen dioxide-induced eosinophilia and mucosal injury in the trachea of the guinea pig. ORL J Otorhinolaryngol Relat Spec 1993; 55:26–40.

84. Chitano P, Mapp CE, Boniotti A, Papi A, Romano M, Salmona M, Ciaccia A, Fabbri LM. Short-term exposure of rats to nitrogen dioxide (NO$_2$) does not change bronchial smooth muscle response in vitro [Abstract]. Am Rev Respir Dis 1993; 147(4 Part 2):A924.

85. Salmona M, Pagani P, Ferro M, DiStefano A, Saetta M, Fabbri L, Erroi A, Romano M. Neutrophil recruitment and elastase-antielastase release in rat bronchoalveolar lavage fluid (BALF) after acute and subchronic exposure to nitrogen dioxide (NO$_2$) [Abstract]. Am Rev Respir Dis 1993; 147(4 Part 2):A672.

86. Januszkiewick AJ, Snapper JR, Sturgis JW, Rayburn DB, Dodd KT, Phillips YY, Ripple GR, Sharpnack DD, Coulson MM, Bley JA. Pathophysiological responses of sheep to brief high level nitrogen dioxide exposure. Inhalation Toxicol 1992; 4:359–372.

87. Atzori L, Bannenburg G, Corriga AM, Lou YP, Lundberg JM, Ryrfeldt A, Moldeus P. Sulfur dioxide-induced bronchoconstriction via ruthenium red-sensitive activation of sensory nerves. Respiration 1992; 59:272–278.

88. Atzori L, Bannenberg G, Corriga AM, Moldeus P, Ryrfeldt A. Sulfur-dioxide-induced bronchoconstriction in the isolated and ventilated guinea pig lung. Respiration 1992; 59:16–21.

89. Norris AA, Jackson DM. Sulfur-dioxide-induced airway hyperreactivity and pulmonary inflammation in dogs. Agents Actions 1989; 26:360–366.

90. Jackson DM, Eady RP. Acute transient SO$_2$-induced airway hyperreactivity: effects of nedocromil sodium. J Appl Physiol 1988; 65:1119–1124.

91. Department of Health. Advisory Group on the Medical Aspects of Air Pollution Episodes. Second Report: Sulphur Dioxide, Acid Aerosols and Particulates. Chapter 4. Biochemical and Toxicological Effects of Sulphur Dioxide, Acid Aerosols and Particulates. London: HMSO, 1992; 39–57.

92. Department of Health. Advisory Group on the Medical Aspects of Air Pollution Episodes. Second Report: Sulphur Dioxide, Acid Aerosols and Particulates. Appendix 3: Reactions and Deposition of Acid Aerosols in the Respiratory Tract. London: HMSO, 1992; 133–152.

93. Pierson WR, Brachaeczek WW, Truex TJ, Butler JW, Korniski TJ. Sulfate measurements on Allegheny Mountain and the question of atmospheric sulfate in the northeastern United States. Ann NY Acad Sci 1980; 338:145–172.

94. Friedlander SK. Future aerosols of the southwest: implications for fundamental aerosol research. Ann NY Acad Sci 1980; 338:588–598.

95. Gearhart JM, Schlesinger RB. Sulfuric acid-induced airway hyperresponsiveness. Fundam Appl Toxicol 1986; 7:681–689.

96. Kobayashi T, Shinozaki Y. Effects of exposure to sulfuric acid aerosol on airway responsiveness in guinea pigs: concentration and time dependency. J Toxicol Environ Health 1992; 39:261–272.

97. Amdur MO, Chen LC. Furnace-generated acid aerosols: speciation and pulmonary effects. Environ Health Perspect 1989; 79:147–150.

98. El-Fawal HAN, Schlesinger RB. Nonspecific airway hyperresponsiveness induced by inhalation exposure to sulfuric acid aerosol: an *in vitro* assessment. Toxicol Appl Pharmacol 1994; 125:70–76.

99. Koenig JQ, Dumler K, Rebolledo V, Williams PV, Pierson WE. The effects of sulfuric acid on nasal lavage and pulmonary function in senior asthmatics [Abstract]. Toxicologist 1992; 12:358.

100. Schlesinger RB, Gorczynski JE, Dennison J, Richards L, Kinney PL, Bosland MC. Long-term intermittent exposure to sulfuric acid aerosol, ozone and their combination: clearance and epithelial secretory cells. Exp Lung Res 1992; 18:505–534.

101. Schlesinger RB, Zelikoff JT, Chen LC, Kinney PL. Assessment of toxicologic interactions resulting from acute inhalation exposure to sulfuric acid and ozone mixtures. Toxicol Appl Pharmacol 1992; 115:183–190.

102. Alarie Y, Busey WM, Kruum AA, Ulrich CE. Long-term continuous exposure to sulfuric acid mist in Cynomolgus monkeys and guinea pigs. Arch Environ Health 1973; 27:16–24.

103. Fujimaki H, Katayama N, Wakamuri K. Enhanced histamine release from lung mast cells of guinea pigs exposed to sulfuric acid aerosols. Environ Res 1992; 58:117–123.

104. Mannix RC, Phalen RF, Nguyen TN. Effects of sulfuric acid on ferret respiratory tract clearance. Inhalation Toxicol 1991; 3:277–291.

105. Chen LC, Fang PC, Qu Q-S, Fine JM, Schlesinger RB. A novel system for the *in vitro* exposure of pulmonary cells to acid sulfate aerosols. Fundam Appl Toxicol 1993; 20:170–176.

106. Zelikoff JT, Schlesinger RB. Modulation of pulmonary immune defense mechanisms by sulfuric acid: effects on macrophage-derived tumor necrosis factor and superoxide. Toxicology 1992; 76:271–281.

107. Chen LC, Fine JM, Qu Q-S, Amdur MO, Gordon T. Effects of fine and ultrafine sulfuric acid aerosols in guinea pigs: alterations in alveolar macrophage function and intracellular pH. Toxicol Appl Pharmacol 1992; 113:109–117.

108. Department of Health. Committee on the Medical Effects of Air Pollutants. Non-Biological Particles and Health. London: HMSO, 1995. [in press]

109. Hinds WC. Aerosol Technology: Properties, behaviour and measurement of airborne particles. John Wiley and Son, 1982.

110. Geiser M, Baumann M, Waber U, Gehr P. Changes in airway macrophage number and phagocytic activity upon particle inhalation [Abstract]. Am Rev Respir Dis 1993; 147(4 Part 2):A13.

111. Oberdörster G, Ferin J, Gelein R, Soderholm SC, Finkelstein J. Role of the alveolar macrophage in lung injury: studies with ultrafine particles. Environ Health Perspect 1992; 97:193–199.

112. Seaton A, MacNee W, Donaldson K, Godden D. Particulate air pollution and acute health effects. Lancet 1995; 345:176–178.

113. Matsumara Y. The effects of ozone, nitrogen dioxide and sulfur dioxide on the experimentally induced allergic respiratory disorder in guinea pigs. I. The effect on sensitization with albumin through the airway. Am Rev Respir Dis 1970; 102:430–437.

114. Matsumara Y. The effects of ozone, nitrogen dioxide and sulfur dioxide on the experimentally induced allergic respiratory disorder in guinea pigs. II. The effects of ozone on the absorption and the retention of antigen in the lung. Am Rev Respir Dis 1970; 102:438–443.

115. Osebold JW, Gershwin LJ, Zee YC. Studies on the enhancement of allergic lung sensitization by inhalation of ozone and sulfuric acid aerosols. J Environ Pathol Toxicol 1980; 3:221–234.

116. Riedel F, Naujukat S, Ruschoff J, Petzoldt S, Rieger CHL. SO_2-induced enhancement of inhalative allergic sensitization: inhibition by anti-inflammatory treatment. Int Arch Allergy Immunol 1992; 98:386–391.

117. Yeadon M, Payne AN. Ozone-induced bronchial hyperreactivity to histamine and ovalbumin in sensitized guinea pigs: differences between intravenous and aerosol challenge [Abstract]. Eur Respir J 1989; 2:299S.

118. Webber SE, Hardman T, Battram C. Combined allergen challenge and ozone exposure produces hyperreactivity to 5HT and airway eosinophilia in guinea pigs *in vivo* [Abstract]. Am Rev Respir Dis 1993; 147(4 Part 2):A785.

119. Kleeberger ST, Turner CR, Kolbe J, Spannhake EW. Exposure to 1 ppm ozone attenuates the immediate antigenic response of canine peripheral airways. J Toxicol Environ Health 1989; 28:349–362.

120. Turner CR, Kleeberger ST, Spannhake EW. Preexposure to ozone blocks the antigen-induced late asthmatic response of the canine peripheral airways. J Toxicol Environ Health 1989; 28:363–371.

121. Bice DE, Mauderly JL, Jones RK, McClellan RO. Effects of inhaled diesel exhaust on immune responses after lung immunisation. Fundam Appl Toxicol 1985; 5:1075–1086.

122. Kaltreider HB, Kyselka L, Salmon SE. Immunology of the lower respiratory tract. II. The plaque-forming response of canine lymphoid tissues to sheep erythrocytes after intrapulmonary or intravenous immunisation. J Clin Invest 1974; 54:263–270.

123. Takafuji S, Suzuki S, Koizumi K, Tadokoro K, Miyamoto T, Ikemori R, Muranaka M. Diesel-exhaust particles inoculated by the intranasal route have an adjuvant activity for IgE production in mice. J Allergy Clin Immunol 1987; 79:639–645.

124. Diaz-Sanchez D, Dotson AR, Takenaka H, Saxon A. Diesel exhaust particles induce local IgE production *in vivo* and alter the pattern of IgE messenger RNA isoforms. J Clin Invest 1994; 94:1417–1425.

125. Nadziejko CE, Nansen L, Mannix RC, Kleinman MT, Phalen RF. Effect of nitric acid vapor on the response to inhaled ozone. Inhalation Toxicol 1992; 4:343–358.

126. Chen LC, Miller PD, Amdur MO, Gordon T. Airway hyperresponsiveness in guinea pigs exposed to acid-coated ultrafine particles. J Toxicol Environ Health 1992; 35:165–174.

127. Chen LC, Miller PD, Lam HF, Guty J, Amdur MO. Sulfuric acid-layered ultrafine particles potentiate ozone-induced airway injury. J Toxicol Environ Health 1991; 34:337–352.

Annex 4A

Absorption of inhaled gases

A4.1   The lung is the major route of uptake of gases and vapours. It is conventional to divide inhaled gases into those which simply equilibrate across the lung and those which react with components of the respiratory system. Modelling of uptake of non-reactive, equilibrating gases, is well advanced and detailed accounts providing access to the original literature are available.[1]

A4.2   The uptake of equilibrating gases is determined by three factors:

the solubility of the gas in blood;

the cardiac output;

the concentration difference between the alveolar space and venous blood.

A4.3   The last of these factors is dependent upon the inhaled concentration of the gas and the ventilation rate. Additionally, a gas such as carbon monoxide which binds avidly to haemoglobin, will equilibrate only slowly as the partial pressure of the gas in blood increases itself only slowly; nitrous oxide, on the other hand, equilibrates rapidly.[2]

A4.4   In the field of air pollution toxicology, the work of Menzel, Miller, Overton, Gerrity and Pryor on the modelling of ozone uptake in the lower respiratory tract has been prominent.[3-8] Experimental work using radioactive isotopes and stable isotopes, eg $^{18}O_3$ has recently been undertaken.

A4.5   The following factors are important in controlling the uptake of reactive gases in the respiratory system:

morphology of the respiratory system including the local morphology of the different regions modelled;

physico-chemical properties of the liquid lining layer in the different regions;

physico-chemical properties of the tissues and blood;

the pattern of breathing: nasal/oral, oral;

ventilatory rate and tidal volume;

physico-chemical properties of the gas;

convective and diffusional patterns of the gas in the system.

A4.6   A great deal of detailed information is, therefore, needed before realistic models of the patterns of absorption can be developed. Ozone, for example, is a highly reactive molecule and reacts with components of the liquid lining layer of the respiratory tract. This has recently been modelled by Pryor,[8] who suggested that ozone molecules could not cross a lung lining fluid layer of greater thickness than 0.1 μm without reacting with components of the layer. The thickness of the lung lining layer varies from one part of the lung to another: in the upper airways it is of the order of 20 μm, in the alveoli, the layer may be only 0.1 μm, possibly as little as 0.01 μm.

A4.7   The variation in thickness is, in part, due to the variation in the thickness of the mucus layer and in part due to variations in thickness of the fluid hypophase. In the ciliated regions of the airway transport of mucus depends upon there being an appropriate liquid hypophase present to allow normal functioning of the cilia. Cilia are of the order of 5–7 μm in length and transient engagement of their apical claws in the undersurface of the mucus, on the forward propulsion stroke, is thought to be necessary for mucus transport. The laterally deviated, recovery stroke takes place in the hypophase. At the transition from the ciliated airway to the unciliated airway, ie, at the junction between the terminal and respiratory bronchioles, there is thought to be a sudden reduction in the thickness of the lining fluid layer. Mucus secretion in the bronchioles is dependent upon surface goblet cells, the submucosal glands being lost with the cartilage plaques on transition from bronchi to bronchioles at about generation 11.

A4.8   It should be noted that there are marked species differences in the pattern of transport of mucus in the airways. In man, a continuous moving sheet of mucus is often assumed in the upper airways. In the rat, on the other hand, mucus may move as distinct streams or as discrete particles.

A4.9   The absorption of gases by the respiratory tract is influenced by the mode of breathing. Nasal breathing exposes inspired gas to a large absorptive surface in the nose and nasopharynx and the dose delivered to the lung is generally less than that which occurs during oral breathing. This is important during exercise: the amount of gas reaching the lung may be increased to a greater extent than that predicted by the change in minute-volume alone.

A4.10   A high percentage of the inhaled mass of common air pollutants: ozone, nitrogen dioxide and sulphur dioxide is removed in the respiratory tract. In man, absorption of sulphur dioxide is almost complete in the upper airways, nose and nasopharynx. Some 80–90% of inspired ozone is extracted by the whole system.[6] In rats, the overall uptake of ozone is about 50% of that inspired.[5]

A4.11   Of the common air pollutants, sulphur dioxide is the best absorbed in the nose and upper airways: in fact, little inspired sulphur dioxide is generally believed to reach the gas exchange of the lung. This view has been questioned by Strandberg[9] who demonstrated, in animal and physical models, that the percentage absorption in the nose was dependent upon the inspired concentration. At low concentrations, absorption in the nose fell to 5% of inspired gas. The reasons for this variation are not clear. Studies in man have shown high absorption rates in the nose.[10] For ozone and nitrogen dioxide, extraction in the nose is less efficient: about 40% of the inspired gas being absorbed there.

A4.12   In modelling the absorption of reactive gases such as ozone, assumptions have been made not only about the distribution of lining fluid in the airways but of the dimensions and branching patterns of the airways themselves. Though the detailed structure of the lung is well understood in man and rat, other species have been less well studied. Mathematical models of branching patterns have to be included in the models of uptake of reactive gases and the use of different anatomical models leads to differences in the predictions of patterns of uptake.

A4.13   Models by different authors of the pattern of uptake of ozone in the lung in a range of species are in general agreement. In terms of net uptake (measures as mass of gas taken up by unit area of surface per minute per μg ozone in the ambient air, ie, uptake by tissue and liquid lining layer), uptake is greatest in the trachea and declines towards the terminal bronchioles. After the terminal bronchioles there is an increase in the net uptake and then a rapid decline as the alveolar ducts and alveoli are approached. This is clearly shown in Figure A4A. 1.

A4.14   The difference between tissue dose and net dose is also shown in Figure A4A.1. It will be seen that the tissue dose is low in the trachea and rises to a peak in the distal airways. In the distal airways and to an even larger extent in the alveoli, the tissue dose approximates the net dose because of the thinness of the liquid lining layer. It has also been suggested that surfactant and perhaps the lining fluid more generally, lacks components capable of reacting with ozone. This should be confirmed experimentally. The precise results of modelling studies of the deposition of ozone in the lung are dependent upon the models of lung structure. Figure A4A. 1 shows the variations in result produced by the use of different structural models.

Figure A4A.1   **Effect of four different LRT anatomical models on predicted net and tissue O_3 dose profiles of guinea pig and rat. Dose is the quantity of O_3 reacting with biological constituents per unit of surface area of a zone, order, or generation, per unit of time per tracheal O_3 concentration. Net dose is the sum of the liquid lining, tissue and blood compartment doses. A: Dose versus zone, anatomical model for guinea pig, V_T = 2.53 mL, f = 56.6 bpm. B: Dose versus order, anatomical model for guinea pig, V_T = 2.4 mL, f = 77 bpm. C: Dose versus zone, anatomical model for rat, V_T = 0.7 mL, f = 144 bpm. D: Dose versus generation, anatomical model for the rat, V_T = 1.84 mL, f = 105 bpm. TB, tracheobronchial region (mucus lined); P, pulmonary region (surfactant lined). Multiply figure values by tracheal concentration ($\mu g/m^3$) to obtain dose ($\mu g/cm^2$ per minute)**

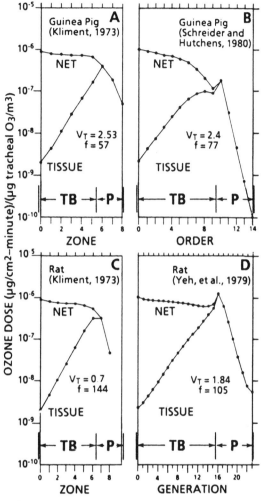

Figure A4A.2   **Net and tissue dose of O_3 versus generation for two humans of different ages (0 and 20 years) subject to two breathing conditions. A, quiet breathing; B, maximal work during exercise. Panels illustrate effect of two extremes of breathing and growth and state on the distribution of absorbed O_3 in the human respiratory tract, given equal ambient O_3 concentrations and no absorption in the URT. NET dose is flux of O_3 to airway or airspace surface. TISSUE dose is rate of absorption of O_3 by epithelial layer in tracheobronchial (TB) region or tissue-blood barrier in pulmonary (P) region. V_T, tidal volume; f, frequency; bpm, breaths min^{-1}**

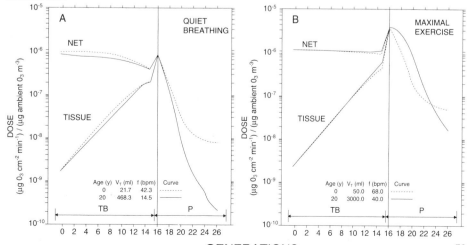

A4.15 Modelling of the uptake of ozone has been extended to models of the human lung: both in the adult lung and in the child's lung. The results obtained are shown in Figure A4A.2 which also shows the effects of different levels of ventilation. It will be seen that exercise increases the dose of ozone received in the distal airway but does not much alter the distribution of the peak local dose.

A4.16 The similarity between the predicted pattern of uptake of ozone in the rat and human is striking. Modelling of uptake patterns in rabbits has also revealed a similar pattern of uptake. Equivalence of local dose should not, however, be assumed between species though the general pattern of uptakes are similar. It has been demonstrated that for exposures of in excess of 100 $\mu g/m^3$ the predicted bronchiolar dose in rabbits was about twice that predicted for guinea pigs and about 80% of that predicted in man. The local dose received by rats, in the lower respiratory tract may be about twice that received by man. Care in extrapolating from data obtained from animal studies, and even more so from the results of modelling studies based upon animal data, to man is clearly needed.

References

1. Hull CJ. The pharmacokinetics of inhalational anaesthetic agents. In: Scurr C, Feldman S, Soni N, editors. Scientific foundations of anaesthesia. 4th Edition. Oxford: Heineman Medical Books, 1990.

2. Comroe JH, Forster FE, DuBois AB, Briscoe WA, Carlsen E. The lung. 2nd Edition. Chicago, London: Year Book Medical Publishers; Lloyd London, 1962.

3. Miller FJ, Menzel DB, Coffin DL. Similarity between man and laboratory animals in regional pulmonary deposition of ozone. Environ Res 1978; 17:84–101.

4. Overton JH, Miller FJ. Modelling ozone absorption in the lower respiratory tract. Presented at the 80th Annual meeting of the Air Pollution Control Association. New York, June 21–26, 1987.

5. Gerrity TR, Weister MJ. Experimental measurements of the uptake of ozone in rats and human subjects. Presented at the 80th Annual meeting of the Air Pollution Control Association. New York, June 21–26, 1987.

6. Overton JH, Graham RC. Predictions of ozone absorption in human lungs from newborn to adult. Health Phys 1989; 57:29–36.

7. Gerrity TR, Weaver RA, Berstsen J, House DE, O'Neil JJ. Extrathoracic and intrathoracic removal of O_3 in tidal-breathing humans. J Appl Physiol 1988; 65:393–400.

8. Pryor WA. How far does ozone penetrate into the pulmonary air/tissue boundary before it reacts? Free Radic Biol Med 1992; 12: 83–88.

9. Strandberg LG. SO_2 absorption in the respiratory tract. Studies on the absorption in rabbits, its dependence on concentration and breathing pace. Arch Environ Health 1964; 9:160–166.

10. Frank NR, Yoder RE, Brain JD, Yokojama E. SO_2 (^{35}S-labeled) absorption by the nose and mouth under conditions of varying concentration and flow. Arch Environ Health 1969; 18:315–322.

Chapter 5

Controlled Chamber Studies

Introduction

5.1 Controlled chamber studies allow investigation of acute airway responses provoked by the inhalation of airborne pollutants. They have been widely used to study the effects of pollutants in normal and asthmatic patients. The benefits of such studies are:

i. The response to inhaled pollutants alone, or in combination with other pollutants or allergen, can be investigated under strictly controlled conditions.

ii. Other factors which may modify the airway response such as temperature, humidity or exercise, can also be strictly controlled.

iii. Identification of exposure-response relationships. They allow accurate measurements of the concentrations of the pollutant within the chamber and the use of well characterised indices of airway response including measures of airway function, such as forced expiratory volume in one second (FEV_1) and airways resistance (Raw), of airway responsiveness, such as the provocation of histamine or methacholine which provokes a 20% fall in FEV_1 (PC_{20}) and the use of fibreoptic bronchoscopy to obtain samples of cells and chemicals within the airways by bronchoalveolar lavage.

iv. Study of different populations permits recognition of "susceptible" groups within the population; and in combination with knowledge of exposure-response relationships, the identification of concentrations of pollutants able to provoke acute airway responses in the "susceptible" groups.

5.2 Chamber studies, however, do have serious limitations:

i. They are limited to the study of acute effects of short-term exposures. Long-term exposures are not feasible and chamber studies are unable to investigate chronic effects.

ii. Those available and selected for study are generally limited to adults with good lung function (and mild asthma). They usually exclude children and those with severe asthma or poor lung function, groups who may be at particular risk from the effects of pollutants.

iii. Although potential interactions between pollutants can be studied, such studies are limited in the number of different pollutants which can be investigated and are unlikely to reproduce natural ambient exposures.

5.3 Despite these limitations, chamber studies can address several of the questions which are important in understanding the relationship of air pollution to asthma:

i. Does a particular pollutant provoke acute airway narrowing in asthmatic patients? If so, what is the nature of the exposure-response relationship and how is this modified by other relevant factors, such as temperature, humidity and exercise?

ii. What is the magnitude of the response in the most sensitive individuals to concentrations of the pollutant (or pollutants) under normal and worst ambient conditions?

iii. Is there any potentiating effect between pollutants or do they potentiate the effect of inhaled allergen on the airways?

Pollutant studies

5.4   The great majority of chamber studies of air pollutants in asthmatic patients have investigated the effects of sulphur dioxide, ozone and oxides of nitrogen. In addition, acid aerosols, particularly sulphuric acid, have also been investigated, as have interactions between these pollutants and of ozone on allergen inhalation.

Sulphur dioxide

5.5   Inhalation of high concentrations of sulphur dioxide (SO_2) provokes acute airway narrowing in normal subjects;[1] lower concentrations provoke the same response in asthmatic patients. In both normal subjects and asthmatic patients, the response is rapid in onset, with a maximum effect in 5 minutes, and resolves within 15 to 30 minutes. The magnitude of response is dependent upon the dose of SO_2 which penetrates to the lower airways; it is, therefore, augmented by oral breathing and increased minute ventilation. Similarly, the response to SO_2 is increased by exercise and hyperventilation of cold, dry air, which are themselves able to provoke asthma. Repeated exposures to SO_2 induce tolerance.

5.6   Although there is considerable variation in response to SO_2 between normal individuals, acute airway narrowing has been provoked by exposure at rest for 10 minutes to airborne concentrations of up to 4,000 and 5,000 ppb (11,440 and 14,300 $\mu g/m^3$),[1] but no important response has been reported below 1,000 ppb (2,860 $\mu g/m^3$), even for prolonged periods (4 hours to 750 ppb, 2,145 $\mu g/m^3$) accompanied by intermittent heavy exercise.[2]

5.7   The nature of the airway response provoked by inhaled SO_2 in asthmatic patients is similar, but provoked by much lower concentrations. There is also considerable variation in response between asthmatic patients, which seems unrelated to the degree of airway hyperresponsiveness (histamine or methacholine PC_{20}) or severity of asthma. In a study of SO_2 exposure-response relationships, Horstman reported a wide variation in the concentration of SO_2 inhaled by male asthmatic patients during moderate exercise which provoked a doubling of airway resistance. The concentration ranged from a minimum of 280 ppb (801 $\mu g/m^3$) to a maximum of 1,900 ppb (5,434 $\mu g/m^3$), with a median value of 750 ppb (2145 $\mu g/m^3$).[3] A small number of asthmatic patients particularly sensitive to SO_2 may have detectable changes in lung function provoked by SO_2 in ambient concentrations of 200 ppb (572 $\mu g/m^3$) during exercise.[4] This compares with ambient hourly concentrations of SO_2 in the UK of usually less than 120 ppb (343 $\mu g/m^3$). On occasions, 200 ppb (572 $\mu g/m^3$) is exceeded in the UK, suggesting that the concentration of SO_2 does, on occasion, exceed the level needed to provoke a fall in lung function in the most sensitive asthmatic patients when undertaking moderate exercise.

Ozone

5.8   The effects of inhaled ozone on the airways have been extensively investigated in controlled chamber studies. The majority of the studies have examined the effects of increasing exposure (concentration and duration) to ozone in individuals undergoing exercise. Inhaled ozone provokes a dose-dependent fall in lung function (reduced FEV_1 and FVC and increased Raw) and an increased, about a doubling in PC_{20}, in airway responsiveness to inhaled histamine. These responses vary considerably between individuals but, surprisingly, little difference has been observed in responses between asthmatic patients and normal individuals.

5.9   The response to ozone is dependent upon the total dose inhaled (intensity and duration of exposure and minute ventilation). The importance of concentration and duration of exposure has been studied by Larsen *et al.*[5] These studies showed that the response to ozone, expressed in terms of decrement in FEV_1, was related to the product of the duration of the exposure and the concentration raised to the power 1.320 (ie, proportional to $t.c^{1.320}$). Usually, the changes in lung function resolve within 24 hours. Repeated daily exposure to ozone stimulates adaptation of shortest duration in individuals most responsive to ozone. Repeated exposure to ozone on 5 successive days provoked a reduction in lung function which was maximal on day 2, with a subsequent daily reduction in response, with a very small effect on day 5.[6]

5.10   The variation in response to ozone in normal individuals was studied by Horstman *et al* in 21 individuals. The fall in FEV_1, after exposure to 120 ppb (240 $\mu g/m^3$) for 6.6 hours ranged from 0 (6 individuals) to 37%. Individual responses were consistent at the two other concentrations of ozone studied (80 and 100 ppb; 160 and 200 $\mu g/m^3$).[7]

5.11   Exposure to ozone in concentrations of between 80 and 120 ppb (160 and 240 $\mu g/m^3$) has been found to increase airway responsiveness in normal individuals. The increase in airway responsiveness is associated in humans with an increase in the proportion of neutrophils recovered at bronchoalveolar lavage, suggesting that ozone provoked an acute inflammatory reaction in the airways, which was present 18 hours after exposure to 80 ppb (160 $\mu g/m^3$) ozone for 6.6 hours.[8]

5.12   Ozone in ambient concentrations increased the fall in FEV_1 provoked by SO_2 in a group of asthmatic patients, but not in a group of normal subjects. Prior inhalation of 120 ppb (240 $\mu g/m^3$) ozone for 45 minutes, by a group of individuals in whom it did not provoke a fall in FEV_1, was associated with a significantly greater reduction in FEV_1 provoked by the inhalation of 100 ppb SO_2, than a similar exposure to SO_2 preceded by clean air.[9] In addition, inhalation of 120 ppb (240 $\mu g/m^3$) ozone for one hour prior to inhalation of ragweed pollen extract was followed by a halving of the concentration of the allergen extract needed to provoke a 20% fall in FEV_1.[10]

5.13   The findings from these studies suggest that ozone, in ambient concentrations, can provoke airway responses in sensitive individuals when undertaking moderate and severe exercise; but, to-date, there is no convincing evidence that asthmatic patients are more responsive to ozone than non-asthmatic patients. However, on the limited evidence available to-date, ozone in ambient concentrations does seem able to augment the effect of SO_2 and inhaled allergen in asthmatic patients but not in non-asthmatic patients.

Oxides of nitrogen   5.14   Inhalation of NO_2 in controlled chamber studies has usually caused no detectable change in measures of lung function in normal or asthmatic patients. The main effect has been a small increase in airway resistance, but this has not been observed consistently. Two studies[11,12] found no change in lung function in asthmatic patients after exposure to NO_2 concentrations of 3,000 and 4,000 ppb (5,640 and 7,520 $\mu g/m^3$) and no change in response to exercise or cold air in the 3,000 ppb (5,640 $\mu g/m^3$) study.[12]

5.15   Orehek reported an increase in airway responsiveness of one doubling dose of carbachol in 13 of 20 asthmatic patients exposed to 100 ppb (188 $\mu g/m^3$) NO_2.[13] However, several subsequent reports failed to find any change in lung function in asthmatic patients following exposure to NO_2 in concentrations up to 1,000 ppb (1,880 $\mu g/m^3$), with and without exercise.[12] Of the studies which investigated the effects of NO_2 between 100 and 1,000 ppb (188 and 1,880 $\mu g/m^3$) on airway responsiveness, two-thirds showed no change, and one-third a small increase. In general, these increases occurred in asthmatic patients when studied at rest, rather than when undertaking exercise and when directly acting stimuli, such as methacholine and histamine were used, rather than indirectly acting agents such as exercise or eucapnic hyperventilation.

5.16   One study found that NO_2 inhaled prior to SO_2 potentiated the bronchoconstrictor response to SO_2. NO_2 when inhaled in a concentration of 250 ppb (470 $\mu g/m^3$) for 30 minutes prior to SO_2 inhalation, augmented the airway response to SO_2, but had no effect on resting lung function.[14] However, another study found that 300 ppb (564 $\mu g/m^3$) NO_2 inhaled by asthmatic patients for 30 minutes had no effect on the response to subsequent inhalations of SO_2 in concentrations up to 4,000 ppb (11,440 $\mu g/m^3$).[15]

5.17   Three recently reported studies[16-18] have provided evidence that NO_2 inhaled prior to inhalation of allergen (in 2 studies house dust mite, in the third pollen) amplified the subsequent asthmatic response to allergen. In one study, a significant increase in both the immediate and late asthmatic response was observed after inhalation of 400 ppb (752 $\mu g/m^3$), but not 100 ppb (188 $\mu g/m^3$) NO_2, as compared to air.[16] In the second study, a decrease in the dose of *D. pteronyssinus* extract which provoked an immediate 20% fall in FEV_1 (PD_{20}) after prior inhalation of 400 ppb (752 $\mu g/m^3$) NO_2 with 200 ppb (572 $\mu g/m^3$) SO_2.[17] In the third study, 500 ppb (940 $\mu g/m^3$) NO_2 prior to pollen inhalation enhanced the magnitude of the late but not immediate asthmatic response.[18]

5.18   The results of the chamber studies of NO_2 in asthmatic patients are inconsistent, but do not suggest that NO_2 inhaled in concentrations in excess of peak concentrations experienced in the UK provokes important changes in lung function or airway responsiveness. More recent evidence suggests that NO_2 at concentrations of 400 ppb (752 $\mu g/m^3$) may enhance the asthmatic response to inhaled allergens. Whether NO_2 potentiates the effect of SO_2 is, as yet, unknown.

References

1.   Nadel JA, Salem H, Tamplin B, Tokiwa Y. Mechanisms of bronchoconstriction during inhalation of sulfur dioxide. J Appl Physiol 1965; 20:164-167.

2.   Stacy RW, Seal E, House DE, Green J, Rodger LJ, Raggio L. A survey of effects of gaseous and aerosol pollutants on pulmonary function of normal males. Arch Environ Health 1983; 38:104-115.

3.   Horstman D, Rodger LJ, Kehrl H, Hazucha M. Airway sensitivity of asthmatics to sulfur dioxide. Toxicol Ind Health 1986; 2:289-298.

4.   Sheppard D, Saisho A, Nadel JA, Boushey AJ. Exercise increases sulfur dioxide induced bronchoconstriction in asthmatic subjects. Am Rev Respir Dis 1981; 123:486-491.

5.   Larsen RI, McDonnell WF, Horstman DH, Folinsbee LJ. An air quality data analysis system for interrelating effects, standards and needed source reductions. Part II: A log normal model relating human lung function decrease to O_3 exposure. J Air Waste Manage Assoc 1991; 41:455-459.

6.   Horvath SM, Gliner JA, Folinsbee LJ. Adaptation to ozone: duration of effect. Am Rev Respir Dis 1981; 123:496-499.

7.   Horstman DH, Folinsbee LJ, Ives PJ, Abdul-Salaam S, McDonnell WF. Ozone concentration and pulmonary response relationship for 6.6 hour exposures with 5 hours of moderate exercise to 0.08, 0.10 and 0.12 ppm. Am Rev Respir Dis 1990; 142:1158-1163.

8.   Holtzman MJ, Fabbri LM, O'Byrne PM, Gold BD, Aizawa H, Walters EH, Alpert SE, Nadel JA. Importance of airway inflammation for hyper-responsiveness induced by ozone. Am Rev Respir Dis 1983; 127:686-690.

9.   Koenig JQ, Covert DS, Hanley QS, van Belle G, Pierson WE. Prior exposure to ozone potentiates subsequent response to sulfur dioxide in adolescent asthmatic subjects. Am Rev Respir Dis 1990; 141:377-380.

10. Molfino NA, Wright FC, Katz I, Tarlo S, Silverman F, McClean PA, Szalai JP, Raizenne M, Slutsky AS, Zamel N. Effect of low concentrations of ozone on inhaled allergen responses in asthmatic subjects. Lancet 1991; 338:199-203.

11. Linn WS, Solomon JC, Trim SC, Spier CE, Shamoo DA, Venet TG, Avol EL, Hackney JD. Effects of exposure to 4 ppm nitrogen dioxide in healthy and asthmatic volunteers. Arch Environ Health 1985; 40:234-239.

12. Linn WS, Shamoo DA, Avol EL, Whynot GD, Anderson KR, Venet TG, Hackney JD. Dose response study of asthmatic volunteers exposed to nitrogen dioxide during intermittent exercise. Arch Environ Health 1986; 41:292-296.

13. Orehek J, Massari JP, Gayrard P, Grimaud C, Sharpin J. Effect of short term low level nitrogen dioxide exposure on bronchial sensitivity of asthmatic patients. J Clin Invest 1976; 57:301-307.

14. Jörres R, Magnussen H. Airways response of asthmatics after a 30 minute exposure at resting ventilation to 0.25 ppm NO_2 or 0.5 ppm SO_2. Eur Respir J 1990; 3:132-137.

15. Rubinstein I, Bigby BG, Reiss TF, Boushey HA. Short term exposure to 0.3 ppm nitrogen dioxide does not potentiate airway responsiveness to sulfur dioxide in asthmatic patients. Am Rev Respir Dis 1990; 141:381-385.

16. Tunnicliffe WS, Burge PS, Ayres JG. Effect of domestic concentration of nitrogen dioxide on airway response to inhaled allergen in asthmatic patients. Lancet 1994; 344:1733-1736.

17. Devalia JL, Rusznak C, Herdman ML, Trigg CJ, Davies RJ. Effect of nitrogen dioxide and sulphur dioxide on airway response of mild asthmatic patients to allergen inhalation. Lancet 1994; 344:1668-1671.

18. Strand V, Svartengen M, Rak S, Bylin G. Effect of NO_2 exposure on an immediate and late response to inhaled allergen in subjects with asthma [Abstract]. Am J Respir Crit Care Med 1994; 149(4 Part 2):A154.

Chapter 6

Panel and Event Studies

Methodological aspects   6.1   To qualify for inclusion in this section, each study must have been designed so that individuals rather than populations were studied. Assessment may have been of either asthmatic patients or normal subjects, but daily recordings of symptoms and/or peak flow will have been made. Studies involving solely patients with COPD have been excluded. Studies involving repeated, though not necessarily daily, measurements of FEV_1 or FVC have been included. This has meant the exclusion of studies such as the Nashville study[1] where weekly data were acquired.

6.2   Although wheeze and breathlessness are the best recognised symptoms of asthma, cough, with or without sputum, is also a cardinal feature of asthma. Where an asthmatic population is studied, therefore, cough and sputum are both relevant and important when considering the potential health effects of pollution. The use of treatment may well have a masking effect on recorded symptoms or peak flow in response to air pollution, particularly in mild asthma. To some extent, this can be assessed by recorded use of treatment which in itself has been used as an outcome variable in some studies, although this should be regarded as a somewhat "soft" outcome variable.

6.3   Panel studies are prospective in nature and usually last for longer than one month. The subjects are selected in advance for the specific purpose of the study. Responses are measured in terms of daily symptoms and peak flow measurements (usually twice daily). Each subject is effectively used as his or her own control which gives small studies the potential to be very sensitive in detecting health effects.

6.4   Event studies may be either prospective, set up in anticipation of a pollution episode (eg, during the winter in Europe, or during ozone summer events in the USA) or may be opportunistic/retrospective studies where a panel of subjects were recording symptoms or peak flow when a pollution event occurred. The majority of these event studies are Summer Camp studies from North America involving, usually, children, and may last for as little as one week. In more recent studies specific attention has been paid to asthmatic children at summer camp. One potential complication of Summer Camp studies is that of tolerance, as may be the case with challenge studies. Children constantly exposed to a certain level of air pollution may show a reduced response to a pollution episode. This may be particularly so if they have not travelled very far to go to camp, and thus the pollution episode may not be unusual in terms of their own experience.

6.5   Methods of analysing data acquired in these ways fall into two main types, analysis by threshold and analysis using continuous variables. Threshold analysis assesses the proportion of an outcome measure (eg, cough) which occurs when pollutant levels exceed a predetermined threshold. This is a simple method but uses only a small amount of the available information. It is also likely to be less sensitive in identifying an effect of small magnitude than analysis using continuous variables where, either for the group under study as a whole, or for each individual, measurement to measurement variation in outcome (eg, symptoms, peak flow) are analysed with respect to pollutant levels. This employs the full range of data, but is a considerably more complex form of analysis.

6.6   Of these methods, the form of statistical analysis now most commonly accepted for such studies is to employ analysis of continuous variables using a regression analysis for each individual, thus deriving a measure of change in some measure of lung function per unit change in pollutant level. This need for a new method of analysing panel data arose from the realisation that associations between a pollutant and a health outcome might be dependent on factors not previously allowed for. As the responses are generally small, other contributing factors, however small, may become crucial. Whittemore and Korn's analysis[2] incorporated the presence or absence of an attack on the previous day, time since start of study and day of the week, as well as meteorological measures. Even so, as a result of their findings, they recommended that in future research, seasonal factors should be allowed for, and that there should be better assessment of individual medication use and details of personal exposures not only to pollutants (particularly environmental tobacco smoke) but also to allergens, viral infections and stress/emotional factors. They also recommended tighter recruitment of patients to exclude those with few or many symptoms. Such a recruitment approach would severely restrict eligible numbers in a given geographical area with a sentinel monitoring site. The approach suggested by Whittemore and Korn is an ideal approach but incorporating all the elements recommended causes huge logistical problems. Indeed it is difficult to assess what effect incorporation of these factors would have on subsequent analyses. The overall effects in themselves are small and, as yet, no studies have specifically assessed personal exposures to pollutants or allergens or the effects of viral infections. It remains open to debate whether personal exposure assessment would strengthen associations or whether non-pollutant exposure would either strengthen or weaken them.

6.7　One advantage of this type of analysis is that the distribution of the regression coefficients can be displayed visually, or a mean calculated. The main problem with this type of analysis is that levels of pollutants and meteorological factors often change together and, in many cases, it is statistically impossible to decide which individual factor is the more important for a given change in health index. One interpretation of this dilemma is that it is incorrect to assume that one particular pollutant may be the sole or major contributor in causing a health effect. It is, in fact, very likely that the mix of pollutants ($\pm$ meteorological factors) is responsible. The role of such interactions are discussed elsewhere in this report and in the 4th report of the Advisory Group on the Medical Aspects of Air Pollution Episodes (MAAPE).[3]

6.8　Autocorrelation, where one day's pollutant levels and health effects are correlated to the values achieved the day before, is a problem in time series data, especially when attempting to assess potential lag effects. Unless accounted for, such an effect will increase statistical significance and thus imply that a lag effect exists where, in fact, it does not. However, lag effects are important in that, when they are consistently and validly shown in different studies, support is lent to the idea that a true causal effect being detected.

Panel studies　　6.9　Over the last 14 years, a number of studies of this type have been undertaken, the majority in the United States.

6.10　Whittemore and Korn's study[2] proved the turning point in analysis as outlined in paragraphs 6.6 and 6.7 above. The major health outcome which emerged from this analysis of 433 patients with asthma (one of the CHESS studies) was that the most important predictor of an individual having an attack was the presence of an attack on the day before. The attacks (defined arbitrarily by each patient) were correlated to both NO_x and SO_2, but not to levels of oxidants.

6.11　Their results also suggested that males were more at risk of showing a response in dry weather and that those subjects who reported more attacks in cold weather were those who were more likely to be affected by pollutant exposure. This suggests that there are specific groups at risk from the effects of air pollution episodes.

6.12　We have considered thirty other panel studies, both prior and subsequent to this paper, published between 1968 and 1994 (See Annex 6A). Most are studies of mixed paediatric populations of subjects with or without asthma. Thirteen include adults in the study population, two of which include patients with COPD.[4,5]

6.13   Three papers are of particular interest in that they divide their paediatric populations into those with and without symptoms, thus avoiding problems with diagnostic labelling.[6,7,8] The study by Vedal et al,[6] of a random population of 351 children, divided their subjects into those with persistent wheeze, those with intermittent wheeze and those who were asymptomatic. No mention was made of cough as a symptom. Seven percent were recorded as having physician diagnosed asthma. No significant correlations were shown between any pollutant and either peak flow or symptoms in any symptom group. The authors advised caution in accepting this as a negative result, as no attempt was made to measure personal exposures.

6.14   The paper by Neas et al from Pennsylvania[7] showed small falls in peak flow when standardised to a specific change in level of pollutant, the changes being seen independently for hydrogen ion, ozone, sulphur dioxide and PM_{10} for the group of 86 as a whole. In those with symptoms, the changes were significantly greater for changes in H^+ and PM_{10}, suggesting that symptomatic children are more susceptible to certain pollutant mixes. In the Utah study from Pope and Dockery[8] of nearly 600 nine to eleven year olds, changes in peak flow correlated negatively with PM_{10} in both the symptomatic and asymptomatic groups. No measure was made of either sulphur dioxide or aerosol strong acid in this study, because levels were known to be low in that area. In an attempt to assess the effects of a spell of poor air quality, five day moving mean values of pollutants were entered into a regression analysis and peak flow changes were found to correlate better with this index than with daily values.

6.15   Apart from those studies which consider symptoms rather than diagnoses, sixteen North American studies[2,4,9-22] have been published, three in normal subjects with no respiratory disease, eight of asthmatic patients and six of normal populations including a proportion of patients with asthma.

6.16   The fourteen studies in which adults have been followed we have divided into studies of patients with asthma and those where normal subjects or random populations samples were used.

6.17   In normal individuals, significant changes were seen in lung function during high ozone days in the New York[9] and Canadian[10] studies, with changes in FEV_1 and FVC of -1.4 and -2.5 ml/ppb increase in ozone, respectively. The only other study in which peak flow was assessed was from Arizona,[11] where small changes in peak flow were seen in normal adults in association with ozone, but this effect was more marked in those with longer outdoor dwell times, in contrast to an earlier study,[4] which had shown no effect in normal subjects. The two Los Angeles studies of nurses from 1974[12] and 1990[13] are contrasting, with the earlier study showing a significant effect on symptoms (27% increase in chest discomfort and 23% increase in cough) with ozone levels above 300 ppb (600 $\mu g/m^3$). The more recent study showed that although oxidants did impact on chest symptoms, passive smoking was more important, with the additional finding that NO_2 exposure was significantly related to phlegm although the associations were weak. The most recent study from California[14] revealed a minute rise (0.44% increase per 10 ppb (18.8 $\mu g/m^3$) rise in NO_2) in symptoms, although this did apparently reach statistical significance. Overall, these findings show an effect of ozone episodes on lung function in normal subjects but only minimal effects on symptoms. The clinical relevance of such short term changes is difficult to assess as are the potential effects of repeated exposures.

6.18   Three studies failed to show any effect of air pollution levels on either symptoms or peak flow in adult asthmatics.[15-17] The earliest of these[15] dates from 1963, before clean air legislation would have had an effect, and studied both adults and children, the predominant pollutant being SO_2. A Californian study,[16] again of children and adults, showed no effects with ozone levels up to 370 ppb (740 $\mu g/m^3$) and SO_2 to 300 ppb (858 $\mu g/m^3$), which may be of more significance than the later study from Denver when pollutant levels were very low (eg, maximum hourly $PM_{2.5}$ 36.5 $\mu g/m^3$).[17]

6.19   Whittemore and Korn's study[2] showed an increase in self-reported asthma attacks by 60% when oxidant levels reached 300 ppb, with a rise of 25% when TSP reached 300 $\mu g/m^3$, while the study from Colorado[18] showed that aerosol strong acid was significantly and positively related to cough, with PM_{10} and sulphate also being related, independently, to cough and breathlessness, respectively. The authors added into their regression an indicator of "outdoor dwell time" which showed that where longer times were spent outdoors effects were increased. The size of this effect was not easily quantifiable in terms of change in symptoms/unit change in pollutant level. Three other studies showed falls in peak flow with increases,[4,11,19] but quantification was again impossible. The study from Utah[20] was unusual in that SO_2 and NO_2 levels were very low, the major pollutant being PM_{10} from a nearby steel mill. Adults and children were analysed together and at a level of 150 $\mu g/m^3$ PM_{10}, falls in peak flow of between 3 and 6% were noted. Overall, adult asthmatic patients show a variable response to air pollution across a range of exposures but results are far from consistent. In particular, there have been no real attempts to quantify any such effects.

6.20   Normal children have been studied as part of a general population sample on five occasions and all five show falls in lung function on exposure to different pollutants. One of the symptomatic/asymptomatic children studies[7] showed an effect of H^+ (1.3 l/m fall in peak flow for a 150 nm/m^3 rise) and of PM_{10} (1.2 l/m fall in peak flow for a 20 $\mu g/m^3$ rise). In other studies, significant negative associations with peak flow[8,11] and ozone[10,11] were noted. A study from New Jersey in 1988 produced a quantified estimate of the effect of ozone on lung function with predicted falls at 120 ppb (240 $\mu g/m^3$) of 4.9% in FVC, 7.7% in FEV_1 and 17% in peak flow.[21] One slightly curious finding was that lung function changes appeared to be more noticeable when ozone levels were less than 60 ppb ($<$ 120 $\mu g/m^3$) compared with less than 80 ppb ($<$ 160 $\mu g/m^3$). Overall, these studies show that lung function in normal children is affected by US summer episodes (as well as adults) although lung function appears to be more affected than symptoms.

6.21   Of those studies which have looked at asthmatic children, two have shown no effect on lung function or symptoms[15,16] as discussed in paragraph 6.18 above. Three have shown significant negative associations with PM_{10}.[7,8,20] Two of these[7,8] are studies of symptomatic vs asymptomatic children as discussed in paragraph 6.13 above, but it is unlikely that the majority of the symptomatic children would have had physician diagnosed asthma. The Utah study,[20] important because of the low levels of other pollutants, showed an effect of PM_{10} on peak flow, which seemed to be better related to a moving 5 day mean PM_{10}, suggesting longer lasting effects of such changes in lung function. Four studies have shown an association with ozone exposure, although in two,[11,19] the effect was studied on a mixed population of adults and children. A third study showed an effect of ozone on various lung function parameters (see paragraph 6.20 above) to a greater degree than in asymptomatic individuals. In the symptomatic group, a 150 nmol/m^3 rise in H^+ was associated with a 1.7 l/m fall in peak flow with a 1.6 l/m fall with a 20 $\mu g/m^3$ rise in PM_{10}. A study in African American children[22] in California, showed a significant effect of ozone but not PM_{10} on breathlessness, the effects being more marked in those from low income families and, independently in those with the more severe asthma. These effects were seen at fairly modest levels of ozone, with an hourly maximum of 80 ppb (160 $\mu g/m^3$). These studies show that asthmatic children are more sensitive than normal children to the effects of ambient ozone and PM_{10} and that effects can occur at modest levels of pollution. The size of the effect is small but, particularly with respect to PM_{10}, the effect may last for longer periods than the day of exposure. They also highlight the importance of daily activity patterns in assessing health effects of air pollution.

6.22   There have been eleven other panel studies from outside the USA and Canada, nine from Europe and one each from Mexico and Australia.

6.23   Romieu and colleagues[23] studied children in Mexico City where the mean hourly maximum ozone level was 200 ppb (400 $\mu g/m^3$). When exposures were assessed over three day periods, respiratory symptoms increased by 55% (although the 48% increase in cough was the only individual symptom which changed significantly). Wheeze was unaffected. Peak flow did show a significant negative association with ozone, but the size of the effect was not repeated.

6.24   A one year study of children living near a power station in Australia measured sulphur dioxide and NO_x only, the levels being very low compared with those found in Europe.[24] No effect on symptoms was shown and no measurements of peak flow were made.

6.25   Two Swiss and two Dutch studies of paediatric populations containing a variable number of asthmatic patients have been reported. The first Swiss study[25] of over one thousand 0 to 5 year olds showed that outdoor NO_2 levels did relate to respiratory symptoms even though both indoor and outdoor levels were very low (16-18 ppb, 30-34 $\mu g/m^3$ indoors; and 13-28 ppb 24-53 $\mu g/m^3$ outdoors). The follow up study[26] from this group then failed to confirm this effect, although outdoor NO_2 did relate to the *duration* of symptomatic episodes. However, this study showed that particulate pollution (as TSP) was a significant predictor of episodes of cough and a better predictor of duration of episodes than NO_2. Brunekreef et al,[27] studying a mix of normal individuals and asthmatic patients, estimated that at 200 $\mu g/m^3$ (100 ppb) ozone, PEF would be 5% lower than on the average ozone level days, but it was not clear whether this was only seen in asthmatic patients or whether there was any difference between asthmatic patients and normal subjects. The later study from that group[28] showed small but significant reductions in peak flow over a winter period when the maximum daily PM_{10} reached 180 $\mu g/m^3$. The effect was small with a 0.41 l/m reduction in peak flow per 10 $\mu g/m^3$ rise in PM_{10}. This was matched by significant associations with wheeze and bronchodilator use.

6.26   In the Czech Republic, SO_2 and PM_{10} still reach levels similar to those experienced in the UK in the 1960s, with daily levels of up to 492 $\mu g/m^3$ (172 ppb) and 171 $\mu g/m^3$, respectively, reported in a study of both adults and children with asthma.[29] Adults were completely unaffected by these exposures both in terms of symptoms and peak flow, while children showed significant changes in lung function but not in symptoms. The effects on peak flow were, however, extremely small, with SO_2 levels at 58 $\mu g/m^3$ (20.3 ppb) producing just a 0.57% fall.

6.27   There have been four studies of adults, one of normal subjects and three of patients with asthma. The small sub-set of 13 normal adults from the Norwegian NILU study[30] by Clench-Aas and colleagues showed an association between NO_2 and both URT symptoms and chest tightness, but showed no change in PEF. The work from Manchester[5] is of interest in that bronchial reactivity was measured at the beginning of the study period. Both asthmatic patients and patients with COPD were studied, but no measure of particles was made. They showed that there was a significant negative relationship between peak flow and both SO_2 and ozone, the latter also being demonstrated at lags of 1 and 2 days. Those with greater bronchial reactivity showed greater responses to these pollutants. A study from Sweden[31] is potentially flawed by being a volunteer study conducted in an area where there was a public perception of an environmental problem. It did, however, demonstrate an association between Black Smoke and symptoms in a small panel of asthmatic patients at low ambient levels (maximum 24 hr level 21.4 $\mu g/m^3$). The study from Birmingham, UK, is the first UK study to measure aerosol strong acid and its constituent levels.[32] Studying adult asthmatic patients with a range of severity, group mean analysis shows significant negative associations of both PEF and symptoms with H^+ and with nitric acid, although the effects are very small. During the summer, acid species, ozone and particles all showed significant negative associations with symptoms and PEF, but again of small degree. A ten-fold increase in nitric acid level (much greater than that actually measured) was estimated to result in a 24% rise in symptom score.

6.28   Overall, these studies show effects of a range of air pollutants on PEF or symptoms, particularly in asthmatic patients. The effects on an individual are small. In normal and asthmatic adults, symptoms seem to increase above 300 ppb (600 $\mu g/m^3$) ozone, with some studies showing no effect. At 150 $\mu g/m^3$ PM_{10}, asthmatic patients show 3-6% falls in peak flow while ozone may have a slightly greater effect on spirometric variables, but this effect has only been seen in one study. The possible effects of acid aerosols need to be considered further in adults. Children seem to be more sensitive to the effects of ozone, particularly if asthmatic. Effects can be seen at levels below 80 ppb (160 $\mu g/m^3$) with, from one study, an estimated 7.7% fall in FEV_1 at 120 ppb (240 $\mu g/m^3$). A 20 $\mu g/m^3$ rise in PM_{10} can reduce peak flow by 0.8-1.6 l/m. Interestingly, where air pollution is known to be worse (Mexico and the Czech Republic) responses to variations in air pollution levels produces less of a response, suggesting a degree of tolerance.

6.29   Studies such as these are not designed to consider long term effects of repeated exposures. Few studies have taken into account atopy or the effects of individual allergen exposures and these should be addressed in future studies. In addition, the majority of these studies did not measure levels of a full range of pollutants. Although multiple regression analysis can quantify individual contributions from certain pollutants to a certain extent, the health effects seen occur usually when more than one pollutant is elevated and assessment of specific combinations of pollutants will need to be addressed. This "cocktail effect" may imply that pollutants can act permissively or interactively in producing health effects, some of which interactions may be investigated by Panel methodology.

Event studies

6.30   Unlike the panel studies, these studies nearly always employ a measure of lung function, either the spirometric variables FEV_1 and FVC (favoured in the Summer Camp Studies) or PEF. Occasionally, measures from the expiratory flow volume curve have been assessed. They can be classified as opportunistic studies as they depend on an unusual pollution event having occurred (in retrospective studies) or being very likely to occur (in prospective studies).

6.31   We have considered sixteen studies, ten from North America and six from Europe (See Annex 6B). Twelve have exclusively studied children, only one of which was a study limited to patients with asthma. One Summer Camp study considered both adults and children. One Dutch study assessed normal adults while the two UK studies included only adult patients with asthma.

6.32   The North American Summer Camp studies[33-42] are, effectively, studies of the effects of ozone and H^+, ie, of acid summer haze. In these studies of "normal" populations of school children, the findings are variable with some studies showing no effect on lung function,[40] while others showed a small reduction in PEF (range -1.2 to -7 ml/s/ppb ozone), FEV_1 (-0.28 to -2.5 ml/ppb ozone), FVC (-0.12 to -2.65 ml/ppb ozone) or MMEF (-0.61 to -2.73 ml/sec/ppb ozone). These effects occurred at ozone levels greater than 110 ppb (220 $\mu g/m^3$).[33,34,36,37,39,41,42] In one study from Canada,[35] an acid event with levels reaching 550 nmoles/m³ resulted in mean falls in peak flow of 7% and of FEV_1 of 3.5%, this effect being most marked in those who showed enhanced methacholine reactivity. The one study of asthmatic children[38] showed that treatment-use correlated with H^+ and sulphate levels (after allowing for temperature) but no details of lung function changes were given in the abstract.

6.33   The earliest event study was published in 1975,[43] a large study of normal adults in Holland exposed to a sulphur dioxide and smoke event in 1969. The effects on FEV_1 were significant with a 150 ml difference between the time of exposure and at follow up, although the follow up period was some three years later.

6.34   Three studies also from Holland, of normal school children of primary school age,[44-46] all showed slight decrements in spirometric variables during typical winter pollution episodes. In one study, the decline was significantly worse at two weeks follow up,[46] but in both that study and an earlier one,[44] lung function had returned towards normal by three weeks. In the later study, MMEF was unaffected at the time of the episode, but was significantly reduced at both two and three weeks.

6.35 A study of a modest SO_2 episode (69 $\mu g/m^3$, 24 ppb) showed a significant negative correlation between morning PEF and the previous day's hourly maximum SO_2, in patients with moderate to severe asthma (mean inhaled steroid dose 1,500 $\mu g/day$). No hourly measures of NO_2 or particles were made during this episode.[47]

6.36 Study of a more complex event in December 1992 in Birmingham, UK, when PM_{10}, NO_2 and SO_2 all increased to hourly maxima of 231 $\mu g/m^3$, 207 ppb (389 $\mu g/m^3$) and 231 ppb (661 $\mu g/m^3$), respectively, over a period of 4 days, showed a 15.5 l/m fall in morning peak flow and a 27.3 l/m fall in evening peak flow in a group of severe asthmatic patients. This occurred despite a marked increase in treatment, whereas in a group of milder patients no such changes were seen.[48] There were significant associations with NO_2 and PM_{10} and peak flow lagged by four days, but not with sulphur dioxide. This suggests that patients with more severe asthma are more likely to be adversely affected by such episodes, although in the milder group, the effects of concomitant treatment may have masked a small effect.

6.37 Event studies have been almost exclusively of normal children exposed to summer events in America or winter events in Europe, or asthmatic adults exposed to European winter events. The winter episodes seem to have a longer lasting effect than those of summer, but there have been no studies of asthmatic children in the winter or adults in the summer. Sulphur dioxide, even at levels within current EC guidelines still appears to be of some importance in Europe, at least in patients with asthma. The summer events show ozone effects but of small degree and probably of limited duration.

Conclusions

6.38 Most Event and Panel studies have shown effects of air pollution on health in both the normal and asthmatic populations, although more so in asthmatic patients. Ozone can cause changes in lung function in both normal and asthmatic patients, but at lower levels in asthmatic patients. Symptoms, however, in patients with asthma tend not to be affected until the levels reach values significantly greater than the current recommended standard of 50 ppb (100 $\mu g/m^3$) as an 8-hour running average. Winter episodes have an effect both on normal children and asthmatic adults in terms of lung function, which may last for as much as a fortnight after an event. Again, however, a greater exposure is required before changes in symptoms are noted. Very few studies have studied patients with more severe asthma, largely because of the worry that treatment might mask any effect. When more severe patients, assessed either in clinical terms or by bronchial reactivity are studied, it is clear that these patients are more likely to be affected either symptomatically or in terms of changes in indices of lung function.

6.39 The overall size of the effect is small with < 1% changes in peak flow and 1-3% changes in symptoms occurring for changes in pollutant levels of around 30% from baseline.

6.40 Panel studies could be used to test hypotheses but in this area to-date have not been so used. Although studies have tended to concentrate on groups thought to be particularly at risk, such as children and patients with asthma, no true comparative studies of different panels with differing perceived risks have been undertaken. Recent work suggesting that elderly patients without lung disease show little response to ozone-containing air pollution[49] highlights the need for comparative studies of the elderly with and without pre-existing lung disease, or those with and without symptoms. Further studies of asthmatic patients with varying degrees of asthma severity may also help in revealing those populations at increased risk. Both these approaches will help quantify populations at risk when estimating the effect of air pollution on public health.

6.41 Equally, there have been no longitudinal panel studies which might have explored changes with time in response to exposures in specific groups, particularly children as their lungs develop.

6.42 Panel studies can also be used to test specific hypotheses with respect to lag effects and the effects of predetermined pollutant mixes on lung function or symptoms.

6.43 It is not at all clear what proportion of a given population might be affected by an air pollution episode though this is important when trying to assess at risk groups and overall effects on public health.

6.44 Winter SO_2/PM_{10}/NO_2 air pollution shows similar effects on patients with asthma but there is less information to suggest that such episodes affect normal subjects, as virtually all studies have considered patients with asthma. Information is needed concerning the effects of winter pollution episodes on normal individuals.

6.45 Future studies should attempt to take into consideration atopy and allergen exposures as well as time/activity analysis to estimate the relative contribution of the indoor and outdoor environments.

References

1. Landau E, Prindle RA, Ziedberg LD. The Nashville Air Pollution Study. Int J Environ Stud 1971; 2:41-45.

2. Whittemore AS, Korn EL. Asthma and air pollution in the Los Angeles area. Am J Public Health 1980; 70:687-697.

3. Department of Health. Advisory Group on the Medical Apects of Air Pollution Episodes. 4th Report: Health Effects of Exposures to Mixtures of Air Pollutants. London:HMSO 1995 [In press].

4. Lebowitz MD, Collins L, Holberg CJ. Time series analyses of respiratory responses to indoor and outdoor environmental phenomena. Environ Res 1987; 43:332-341.

5. Higgins BG, Francis HC, Warbuton CJ, Yates C, Fletcher AM, Pickering CAC, Woodcock AA. Effects of air pollution on peak expiratory flow measurements in patients with asthma and chronic bronchitis [Abstract]. Thorax 1993; 48:417.

6. Vedal S, Schenker MB, Munoz A, Samet JM, Batterman S, Speizer FE. Daily air pollution effects on children's respiratory symptoms and peak expiratory flow. Am J Public Health 1987; 77:694-698.

7. Neas LM, Dockery JW, Spengler JD, Speizer FE, Tollerud DJ. The association of ambient air pollution with twice daily peak expiratory flow measurements in children [Abstract]. Am Rev Respir Dis 1992; 145(4 Part 2):A429.

8. Pope CA, Dockery DW. Acute health effects of PM_{10} on symptomatic and asymptomatic children. Am Rev Respir Dis 1992; 145:1123-1128.

9. Spektor DM, Lippmann M, Thurston GD, Lioy PJ, Stecko J, O'Connor G, Garshick E, Speizer FE, Hayes C. Effects of ambient ozone on respiratory function in healthy adults exercising outdoors. Am Rev Respir Dis 1988; 138:821-828.

10. Brauer M, Vedal S, Brook J. Effect of ambient ozone exposure on same day lung function change in adult farm workers [Abstract]. Am J Respir Crit Care Med 1994; 149(4 Part 2):A659.

11. Krzyzanowski M, Quackenboss JJ, Lebowitz MD. Relation of peak expiratory flow rates and symptoms to ambient ozone. Arch Environ Health 1992; 47:107-115.

12. Hammer DI, Hasselblad V, Portnoy B, Wehorle PF. Los Angeles student nurse study: daily symptom reporting and photochemical oxidants. Arch Environ Health 1974; 28:255-260.

13. Schwartz J, Zeger S. Passive smoking, air pollution and acute respiratory symptoms in a diary study of student nurses. Am Rev Respir Dis 1990; 141:62-67.

14. Ostro BD, Lipsett MJ, Mann JK, Krupnick A, Harrington W. Air pollution and respiratory morbidity among adults in Southern California. Am J Epidemiol 1993; 137:691-700.

15. Brown EB, Ipsen J. Changes in severity of symptoms of asthma and allergic rhinitis due to air pollutants. J Allergy 1968; 41:254-268.

16. Kurata KH, Glovsky M, Newcomb RL, Easton JG. A multi-factorial study of patients with asthma. Part 2: Air pollution, animal dander and asthma symptoms. Ann Allergy 1976; 37:398-409.

17. Perry GB, Chai H, Dickey DW, Jones RH, Kinsman RA, Morrill CG, Spector SL, Weiser PC. Effects of particulate air pollution on asthmatics. Am J Public Health 1983; 73:50-56.

18. Ostro BD, Lipsett MJ, Weiner MB, Selner JC. Asthmatic responses to airborne acid aerosols. Am J Public Health 1991; 81:694-702.

19. Lebowitz MD, Holberg CJ, Boyer B, Hayes C. Respiratory symptoms and peak flow associated with indoor and outdoor air pollutants in the Southwest. JAPCA 1985; 35:1154-1158.

20. Pope CA, Dockery DW, Spengler JD, Raizenne ME. Respiratory health and PM_{10} pollution. Am Rev Respir Dis 1991; 144:668-674.

21. Spektor DM, Lippmann M, Lioy PJ, Thurston GD, Citak K, James DJ, Bock N, Speizer FE, Hayes C. Effect of ambient ozone on respiratory function in active, normal children. Am Rev Respir Dis 1988; 137:313-320.

22. Ostro B, Lipsett M, Mann J, Braxton-Owens H. An epidemiologic investigation of the effects of air pollution on pediatric asthma [Abstract]. Am J Respir Crit Care Med 1994; 149(4 Part 2):A658.

23. Romieu I, Meneses F, Sienra JJ, Huerta J, Ruiz-Velasco S, White M, Etzel R, Hernandez-Avila M. Effect of ambient ozone on respiratory health in Mexican children with mild asthma [Abstract]. Am J Respir Crit Care Med 1994; 149:(4 Part 2):A659.

24. Henry RL, Bridgman HA, Wlodarcyzyk J, Abramson R, Adler JA, Hensley MJ. Asthma in the vicinity of power stations: II. Outdoor air quality and symptoms. Pediatr Pulmonol 1991; 11:134-140.

25. Rutishauser M, Ackermann-Liebrich U, Braun CH, Gnehm HP, Wanner HU. Significant association between outdoor NO_2 and respiratory symptoms in preschool children. Lung 1990; 168(Suppl):347-352.

26. Braun-Fährlander C, Ackermann-Liebrich U, Schwartz J, Gnehm HP, Rutishauser M, Wanner HU. Air pollution and respiratory symptoms in preschool children. Am Rev Respir Dis 1992; 145:42-47.

27. Brunekreef B, Lebret E, Hoek G, van Kessel A. Effects of ozone on lung function in children living in the Netherlands [Abstract]. Arch Environ Health 1991; 46:119.

28. Roemer W, Hoek G, Brunekreef B. Effect of ambient winter air pollution on respiratory health of children with chronic respiratory symptoms. Am Rev Respir Dis 1993; 147:118-124.

29. Peters A, Beyer U, Spix C, Kranke F, Gutschmidt K, Heinrich J, Dockery D, Dumyahn T, Spengler J, Vhylidahl F, Schubertova E, Wichmann HE. Short term effects of air pollution in a polluted area in the Czech Republic [Abstract]. Am J Respir Crit Care Med 1994; 149(4 Part 2):A662.

30. Clench-Aas J, Larssen S, Bartonova A, Aarnes MJ, Myhre K, Christensen CC, Neslein IL, Thomassen Y, Levy F. The health effects of traffic pollution as measured in the Vålerenga area of Oslo, Lillestrom. Summary report. Lillestrom: Norwegian Institute for Air Research, 1991.

31. Forsberg B, Stjemberg N, Falk M, Lundback B, Wall S. Air pollution levels, meteorological conditions and asthma symptoms. Eur Respir J 1993; 6:1109-1115.

32. Walters SM, Ayres JG, Archer G, Harrison RM. Effects of aerosol strong acid on respiratory function in asthmatic patients: preliminary data from a panel study [Abstract]. Am J Respir Crit Care Med 1994; 149(4 Part 2):A661.

33. Lippmann M, Lioy PL, Leikauf G, Green KB, Baxter D, Morandi M, Pasternack BS. Effects of ozone on the pulmonary function of children. Advances in Modern Environmental Toxicology. New York: Institute of Environmental Medicine, 1983; 423-446.

34. Lioy PJ, Vollmuth TA, Lippmann M. Persistence of peak flow decrement in children following ozone exposures exceeding the National Ambient Air Quality Standard. JAPCA 1985; 35:1068-1071.

35. Raizenne M, Burnett RT, Stern B, Franklin CA, Spengler JD. Acute lung function responses to ambient acid aerosol exposures in children. Environ Health Perspect 1989; 79:179-185.

36. Berry M, Lioy PJ, Gelperin K, Buckler G, Klolz J. Accumulated exposure to ozone and measurement of health effects in children and counsellors at two summer camps. Environ Res 1991; 54:135-150.

37. Spektor DM, Thurston GD, Mao J, He D, Hayes C, Lippmann M. Effects of single and multi-day ozone exposures on respiratory function in active normal children. Environ Res 1991; 55:107-122.

38. Thurston G, D'Souza N, Lippmann M, Bartoszek N, Fine I. Associations between summer haze air pollution and asthma exacerbations: a pilot camp study [Abstract]. Am Rev Respir Dis 1992; 145(4 Part 2):A429.

39. Dockery DW, Ware JH, Ferris BG, Herman SM. Change in pulmonary function in children associated with air pollution episodes. JAPCA 1982; 32:937-942.

40. Stebbings JH, Fogleman DG, McClain KE, Townsend MC. Effect of the Pittsburgh air pollution episode upon pulmonary function in schoolchildren. JAPCA 1976; 26:547-553.

41. Higgins ITT, D'Arcy JB, Gibbons DI, Avol EL, Gross KB. Effect of exposures to ambient ozone on ventilatory lung function in children. Am Rev Respir Dis 1990; 141:1136-1146.

42. Kinney PL, Ware JH, Spengler JD, Dockery DW, Speizer FE, Ferris BG. Short term pulmonary function change in association with ozone levels. Am Rev Respir Dis 1989; 139:56-61.

43. van der Lende R, Huygen C, Jansen-Koster EJ, Knijpstra S, Peset R, Visser BF, Wolfs EH, Orie NG. A temporary decrease in the ventilatory function of an urban population during an acute increase in air pollution. Bull Physiopath Respir 1975; 11:31-43.

44. Dassen W, Brunekreef B, Hoek G, Hofschreuder P, Staatsen B, de Groot H, Schouten E, Biersteker K. Decline in children's pulmonary function during an air pollution episode. JAPCA 1986; 36:1223-1227.

45. Brunekreef B, Lumers M, Hoek G, Hofschreuder P, Fischer P, Biersteker K. Pulmonary function changes associated with an air pollution episode in January 1987. JAPCA 1989; 39:1444-1447.

46. Hoek G, Brunekreef B, Hofschreuder P, Lumens M. Effects of air pollution episodes on pulmonary function and respiratory symptoms. Toxicol Ind Health 1990; 6:189-197.

47. Ayres JG, Rayfield D, Appleby RM, Giles B. Peak flow assessment during an air pollution episode [Abstract]. Am Rev Respir Dis 1991; 143(4 Part 2):A94.

48. Walters SM, Miles J, Archer G, Ayres JG. Effect of an air pollution episode on respiratory function of patients with asthma [Abstract]. Thorax 1993; 48:1063.

49. Höppe P, Lindner J, Prani G, Bronner N, Fruhmann G. Comparison of lung function parameters of senior citizens on days with low and elevated environmental ozone concentrations [Abstract]. Am J Respir Crit Care Med 1994; 149(4 Part 2):A659.

Annex 6A

Summary of panel studies investigating the effects of air pollutants upon normal or asthmatic patients

Author(s)	Year	Country, study area	Age	Number	Asthmatic or normal patients	Effects
Whittemore and Korn[2]	1980	USA, Los Angeles	<16	259	Asthmatic patients	Strong cross-correlation of pollutants. Presence or absence of symptoms on preceding day was an important predictor. Attacks correlated to both O_3 and TSP. RR 1.6 at 300 ppb O_3; 1.25 at $300 \mu g/m^3$ TSP
Lebowitz et al[4]	1987	USA, Arizona	Adults		Asthmatic patients, COPD, normals	In asthmatic patients, symptoms and PEF related to O_3, NO_3, allergens and meteorogical factors
Higgins et al[5]	1992	UK, Manchester	Adults 18–82	77	Asthmatic patients Chronic bronchitics	37 showed MCh reactivity. PEF negatively related to SO_2 and O_3 with 24 and 48 hour lags for O_3. More marked in those who react
Vedel et al[6]	1987	USA, Chestnut Ridge	Children Mean age 10.1	351	Both (7% asthmatic patients)	Mean maximum hourly NO_2 21.5 ppb. Three groups: persistent wheeze; intermittent wheeze; asymptomatic. No correlations with any pollutant and either symptoms or PEF
Neas et al[7]	1990	USA, Pennsylvania	Children	86	62 symptomatic 23 asymptomatic	PEF fell by 1.3 1/min for a 150 nmole/m^3 increase in H^+. Also changes with O_3 (−1.0 1/min/20 ppm); SO_2 (−0.9 1/min/15 $\mu g/m^3$); PM_{10} (−1.2 1/min/20 $\mu g/m^3$). In the symptomatic group: −1.7 1/min/150 nmole rise in H^+; −1.6 1/min/20 $\mu g/m^3$ rise in PM_{10}
Pope and Dockery[8]	1990/91	USA, Utah	Children 9–11	591	129 symptomatic 462 asymptomatic	No SO_2 measured. H_2SO_4 low (0.5 $\mu g/m^3$). PEF negative correlation with PM_{10} in both groups. Symptoms more affected than PEF. PEF related to 5 day moving mean
Spektor et al[9]	1988	USA, New York	Adults Mean age 33	30	Normal, non smokers	Pre and post-exercise lung function measured. O_3 levels 21–124 ppb (hourly) significant (p < 0.01) negative association between O_3 and lung function: FVC −2.1 ml/ppb; FEV_1 −1.4 ml/ppb; PEF −9.2 ml/s/ppb; FEF_{25-75} −6.0 ml/s/ppb. Effects larger than comparable chamber studies
Brauer et al[10]	1994	Canada, British Columbia	10–69	58	General population	Pre and post shift spirometry over 2 months. Mean hourly ozone 30 ppb. Weighted slopes −2.5 ml (FEV_1) and −3.0 ml (FVC) for each ppb increase in O_3 (p < 0.01)

Author(s)	Year	Country, study area	Age	Number	Asthmatic or normal patients	Effects
Kryzyzanowski et al[11]	1992	USA, Arizona	Children NS adults	287 523	Both	PEF fell significantly on ozone days. More marked in asthmatic patients. Fall in PEF in adults seen in those with longer outside dwell times. Increased effect with increased temperature and PM_{10}
Hammer et al[12]	1974	USA, Los Angeles	Adults 18–22	64	Normal population study	Chest discomfort (+27%) and cough (+23%) increased at 300 ppb O_3. Above 400 ppb symptoms increased by 222% and 77%, respectively. No measure of PEF
Schwartz et al[13]	1990	USA	Nurses	100	Both (5% asthmatic patients)	3 year study. NO_2 significantly related to phlegm, 1 hr max NO_2 110 ppb. Oxidants significantly related to chest discomfort but passive smoking more important
Ostro et al[14]	1993	USA, California	Adults	321	Normal (6% had a "chronic respiratory condition")	LRT symptoms significantly associated with 1 hr ozone max levels (22% increase for a 10 pphm rise). Similar rise for a 10 $\mu g/m^3$ increase in sulphates. Presence of gas stove increased symptom risk by 23%
Brown and Ipsen[15]	1968	USA, Philadelphia	Children and adults	314	Asthmatic patients/ rhinitis	No significant correlations with pollutants and symptoms on multiple regression analysis
Kurata et al[16]	1976	USA, California	Mixed 7–72	64	Asthmatic patients	No association between symptoms and O_3 (up to 370 ppb), SO_2 (300 ppb), NO_2 or PM. PEF not measured
Perry et al[17]	1983	USA, Denver	Adults	24	Asthmatic patients	3 month period with no high particulate days. (Max 12 hour PM_{10} 36.5 $\mu g/m^3$). No significant association with symptoms or PEF (except one isolated result at $p < 0.05$ for "fine nitrates")
Ostro et al[18]	1987/8	USA, Colorado	Mean age 44.3	207	Asthmatic patients	H^+ related to cough and SOB. PM_{10} related to cough, SO_2 related to SOB. Including outdoor dwell times enhanced these associations
Lebowitz et al[19]	1985	USA, Tucson	5 to adult	204	45 asthmatic patients 62 allergic patients 68 AOD patients 29 asymptomatics	Indoor and outdoor assessment made. In 5–27 yr group, PEF correlated negatively to O_3 levels. In asthmatics PEF associated with smoking, gas stove use, outdoor NO_2, outdoor O_3, and outdoor temperature. Unable to estimate size of effects

Author(s)	Year	Country, study area	Age	Number	Asthmatic or normal patients	Effects
Pope et al[20]	1991	USA, Utah	8–9 8–72	41 25	Asthmatic patients Asthmatic patients	Increase in PM_{10} to 150 $\mu g/m^3$ caused 3 to 6% decline in PEF. Very low levels NO_2 and SO_2. Effects accumulated over several days exposures
Spektor et al[21]	1988	USA, New Jersey	8–15	92	Normal patients	Significant ($p < 0.05$) negative associations between various lung function measurements and O_3 levels. FVC –1.03 ml/ppb O_3; FEV_1 –1.42 ml/ppb O_3; PEF –6.78 ml/s/ppb O_3; FEF_{25-75} –2.48 ml/s/ppb O_3
Ostro et al[22]	1994	USA, California	7–13		Asthmatic patients	Maximum hourly O_3 80 ppb and PM_{10} 56 $\mu g/m^3$. Significant association of O_3 but not PM_{10} with breathlessness. Effects more marked in the more severe asthmatics and in low income families
Romieu et al[25]	1994	Mexico	5–13	67	Asthmatic patients	Mean hourly maximum O_3, 200 ppb, PM_{10} 54 $\mu g/m^3$. Analysed by strata. 3 day exposures led to 55% increase in overall symptoms (+48% for cough). Wheeze was not affected
Henry et al[24]	1991	Australia	Children	99	Asthmatic and normal patients	One year study. Low SO_2 levels. No relationship of symptoms with SO_2 or NO_x. But overall symptom prevalence was low
Rutishauser et al[25]	1990	Switzerland	0–5	1225	Asthmatic and normal patients	NO_2 indoor 16–18 ppb. Outdoor 13–28 ppb. NO_2 related to respiratory symptoms (10% increase in symptom score for each 10 $\mu g/m^3$ rise in NO_2)
Braun-Fährlander et al[26]	1992	Switzerland	0–5		Asthmatic and normal patients	Respiratory symptoms related to TSP but not NO_2. Length of symptomatic episodes related to NO_2
Brunekreef et al[27]	1989	Holland	Children	500	Asthmatic and normal patients	Maximum hourly O_3 200 $\mu g/m^3$. Estimated fall in PEF 5% at 200 $\mu g/m^3$ O_3. Spirometry not affected.

Annex 6A continued

Author(s)	Year	Country, study area	Age	Number	Asthmatic or normal patients	Effects
Roemer et al[28]	1993	Holland	Children 6–12	73	Patients with chronic respiratory symptoms (21% asthmatic patients)	Over one winter, included an episode (24 hr $PM_{10} > 110\,\mu g/m^3$, peak $180\,\mu g/m^3$). Small significant reductions in PEF with PM_{10}, BS, SO_2, also related to wheeze and bronchodilator use. No correlation with H^+, NO_2
Peters et al[29]	1994	Czech Republic	Children Adults	94 59	Asthmatic patients 60% asthmatic patients	Daily SO_2: $2–492\,\mu g/m^3$. Daily PM_{10}. $3–171\,\mu g/m^3$. Adults not affected. Children: no symptoms recorded; PEF fell by 0.57% for a $58\,\mu g/m^3 \uparrow$ in SO_2
Clench-Aas et al[30]	1991	Norway	Adults	13	Both	Small subset of main study. Significant, slight relationship between NO_2 and URT symptoms and chest tightness. No relationship with PEF
Forsberg et al[31]	1993	Sweden	Adults	31	Asthmatic patients	Volunteer study. Measured BS, NO_2 and SO_2. BS significantly associated with breathlessness in those with more initial symptoms
Walters et al[32]	1994	UK, Birmingham	17–85	33	Asthmatic patients	Winter data. Lagged association between PEF and HCl, NH_3, NO_2 and SO_4

FEV_1 = forced expiratory volume in 1 second; $FEV_{25–75}$ = the mean forced expiratory flow rate during the middle half of a forced expiration from full inspiration (l/sec); FVC = forced vital capacity; LRT = lower respiratory tract; MCh = methacholine; PEF = peak expiratory flow; RR = relative risk; SOB = shortness of breath; URT = upper respiratory tract

Annex 6B

Summary of event studies investigating the effects of air pollutants on asthma

Author(s)	Year	Country, study area	Age group (n)	Population	Pollutants	Effects
Lippmann et al[3]	1983	USA	8–13 yr (83)	Normal	O_3, TSP, H_2SO_4	At levels of O_3 of 110 ppb significant, but small reductions in FEV_1 and FVC compared to days at 80 ppb (−1.06 ml/ppb FVC; −0.78 ml/ppb FEV_1)
Lioy et al[34]	1985	USA, New Jersey	7–13 yr (62)	Normal	O_3, $PM_{2.5}$, H^+ NO_3, SO_4	Suggested that PEF in girls negatively associated with O_3 but that the initial O_3 event (180 ppb) had a persistent effect over 3 weeks. −3 ml/sec/ppb for PEF; −0.12 ml/ppb FVC
Raizenne et al[35]	1989	Canada, Ontario	8–14 yr girls	Normal	O_3, H^+, TSP, NO_2, SO_2 low	Episodes involving O_3: 143 ppb (peak 1 hr) with max H_2SO_4 47.4 $\mu g/m^3$ and H^+ 55 nmole/m^3 led to mean falls of 3.5% and 7.5% in FEV_1 and PEF, respectively. Effects more marked in those who were MCh sensitive
Berry et al[36]	1991	USA, New Jersey	9–35	Normal	O_3, H^+	In children PEF fell by 4.74 ml/sec/ppb O_3 (8 hr average) persisting over 3 days. Symptoms increased at levels more than 120 ppb
Spektor et al[37]	1991	USA, New Jersey	8–14 yr (46)	Normal	O_3, H^+	Effect of O_3 but not H^+ on FEV_1 and PEF. Previous day effect.
Thurston et al[38]	1991	USA, Connecticut	7–13 yr (50)	Asthmatic patients	O_3, H^+, SO_4	Max 1 hr O_3, 154 ppb. Daytime H^+ 245 nmoles/m^3. Daytime SO_4 26.7 $\mu g/m^3$. Treatment use significantly correlated with H^+, SO_4 allowing for temperature. No measure of lung function given
Dockery et al[39]	1982	USA	7–8 yr (200)	Normal	TSP, SO_2	Median fall in FEV_1 0.75% and FVC of 1% during episodes. 2.1% fall in FVC 2 weeks after episode. 24 hr max TSP 312 $\mu g/m^3$; 24 hr max SO_2 455 $\mu g/m^3$
Stebbings et al[40]	1976	USA, Pittsburgh	8–9 yr (272)	Normal	COH, SO_2, TSP NO_2	Max hourly SO_2 410 ppb. Max hourly TSP 462 $\mu g/m^3$. No changes with $FEV_{0.75}$ or FVC. No correlation with pollutants

Annex 6B continued

Author(s)	Year	Country, study area	Age group (n)	Population	Pollutants	Effects
Higgins et al[41]	1990	USA, California	7–13 yr (43)	Normal	PM_{10}, $PM_{2.5}O_3$, NO_2, SO_2	1 week. No passive cigarette smoke exposure. O_3 max 245 ppb (1 hr). PEF not affected. FEV_1 decline 0.39 ml/ppb O_3. FVC decline 0.44 ml/ppb O_3. Where changes occurred, these reversed quickly with fall in ozone, but effects remained positive with respect to 2 hr ozone lag
Kinney et al[42]	1989	USA	7–13 yr (154)	Normal	O_3	Max O_3 78 ppb (1 hr). Small reduction in spirometry: 0.99 ml/ppb O_3 for $FEV_{0.75}$; 0.92 ml/ppb O_3 FVC; 1.9 ml/s/ppb O_3 for FEF_{25-75}
van der Lende et al[43]	1975	Holland	15–64 yr (3788)	Normal	SO_2, NO_2	Data from 1969 and 1972. Max SO_2 (1969 vs 1972) 300 vs 110 $\mu g/m^3$. Measured over 5 days. 150 ml differences in FEV_1 between years
Dassen et al[44]	1986	Holland	6–11 yr (636)	Normal	SO_2, TSP, NO_2	Small temporary falls in FEV_1 and FVC when SO_2 and TSP reached 200–250 $\mu g/m^3$, returned to normal after 3 weeks
Brunekreef et al[45]	1987	Holland	4–10 yr	Normal	SO_2, TSP, NO_2	Transient falls in spirometry with raised levels of SO_2 and particulates
Hoek et al[46]	1990	Holland	4–10 yr (191)	Normal	SO_2, NO_2, BS, SO_4^{2-}, NO_3^-, TSP	Peak SO_2, 289 $\mu g/m^3$ (daily). Peak TSP 278 $\mu g/m^3$ (daily). Significant reductions in FEV_1 (−17 ml) and FVC (−20 ml) at time of episode, with prolonged effects at 2 weeks (−77 ml and −78 ml) and 3 weeks (−52 ml, −40 ml). MMEF significantly reduced (−91 ml/s and −57 ml/s) at 2 follow up week only
Rayfield et al[47]	1992	UK, Birmingham	20–62	Asthmatics	SO_2	Significant correlation of morning PEF to previous day hourly maximum SO_2. But, NO_2 and BS not measured
Walters et al[48]	1993	UK, Birmingham	19–55	Asthmatic patients, severe, mild	PM_{10}, SO_2, NO_2	Significant correlation of fall in PEF and treatment in the severe, but not the mild group

FVC = Forced vital capacity; FEV_1 = the volume of air expired during the first second of a maximal or "forced" expiration; $FEV_{0.75}$ = the volume of air expired during the first 0.75 seconds of a maximal or "forced" expiration; FEF_{25-75} = the mean forced expiratory flow rate during the middle half of a forced expiration from full inspiration (1/sec)

Chapter 7

Asthma, Allergy and Air Pollution in Great Britain: Time Trends and Geographical Variations

Introduction

7.1 This chapter reviews the evidence concerning time trends and geographical variations in asthma and other allergic diseases in Great Britain, and the corresponding temporal and spatial trends. These include:

a) Long-term time trends in asthma and other allergic diseases in the UK and elsewhere over recent decades, their relationship to trends in air pollution levels and other factors.

b) Seasonal variations in occurrence of asthma in the UK and the relationship of these to variations in air pollution levels and other factors.

c) Regional and urban-rural variations in asthma and other allergic diseases in the UK and selected other countries and the corresponding variations in past and current air pollution levels.

7.2 Short-term (eg, daily) variations in occurrence of asthma attacks and their relationship to air pollution levels will be discussed in Chapter 8. This will include consideration of "asthma epidemic days" and analyses of extended time-series in the UK and elsewhere. The relationships of geographical variations in asthma prevalence or severity to exposure to vehicle emissions, point sources of air pollution, or measured pollutant concentrations in countries other than the UK are discussed in Chapter 9.

Long-term time trends

Asthma: general

7.3 The nature and the possible measures of the occurrence of asthma within populations were introduced in Chapter 3. This section will review, in turn, the evidence relating to time trends in prevalence, health service utilisation and mortality due to asthma in the UK over the past 30 years, summarising evidence from abroad where appropriate. In conclusion, evidence from the various sources will be considered together and apparent discrepancies will be explored.

Asthma: prevalence in Britain

7.4 Defining the prevalence of an episodic complaint such as asthma presents a problem in that most people with a tendency to wheeze are asymptomatic at any given moment. It has become customary, therefore, to refer to period prevalence, defined as the proportion of the population with one or more episodes of wheezing over a specified prior period, usually the last twelve months. The annual period prevalence of wheezing is an attractive index to describe the prevalence of asthma in school children because at this age few other diseases cause wheezing and treatment will rarely be sufficiently effective to render a patient free of all wheezing for as long as twelve months. However, it may be a poor guide to the occurrence of more severe manifestations of asthma, and should be applied with caution in infants and older adults, among whom wheezing may be due to diseases other than asthma.

7.5 Much of our understanding of the long-term time trends in asthma prevalence in the UK is derived from studies of children of school age. Remarkably little is known directly about the prevalence of asthmatic symptoms in pre-school children,[1] but some indication of time trends in this group can be gleaned from parental responses to questions enquiring about lifetime prevalence (*"asthma or wheezing ever"*) in surveys of school children. The following sections will concentrate on the latter age group.

7.6 In general, surveys of British school children which have enquired about specific diagnoses, such as asthma or wheezy bronchitis, obtain somewhat lower prevalence figures, by as much as one-quarter or one-half, than those which refer to symptoms such as wheezing. The wording of the symptom questions may also be critical: about 10% of parents of asthmatic children may not report their child's wheezing if the questions refer to "attacks" or "episodes" of wheeze.

7.7 A compilation of all British studies which have measured annual period prevalences of wheezing (or a broad range of diagnostic categories) suggests that there may have been a modest increase in the period prevalence, perhaps of the order of 50% in relative terms, from the late 1960s to the 1980s.[2] These studies, which are summarised in Table 7.1 (modified from Lenney et al[3]), used a variety of survey techniques, age groups and catchment areas, and it is difficult to draw firm conclusions from such an overview.[4-23] However, it is clear that the prevalence of diagnosed asthma has increased much more rapidly than the prevalence of wheezing and there is a suggestion that the prevalence of wheezing may have reached a plateau during the 1980s (see Table 7.1 and Figures 7.1 and 7.2).

Table 7.1 **Population-based prevalence studies of asthma and wheezing among school-aged children in the UK.** (*Modified and updated from Lenney et al[3]*)

Ref	Year of study	Area	Age	Subjects Number	Wheeze past yr	Recent wheeze	Wheeze ever	Recent asthma	Asthma ever	Type of study
							Prevalence (% of)			
Dawson et al[4]	1964	Aberdeen	10-15	2511		11.5		4.6		General
Graham et al[5]	1964	Isle of Wight	9-11	3300				2.1		Medical
Anderson et al[6]	1965	National	7	14571	8.1		18.3		3.1	General
Hamman et al[7]	1967	Kent	5-14	10971					3.8	Chest
Leeder et al[8]	1968	Harrow	5	2037			24.9		3.2	Chest
Smith et al[9]	1968-9	Birmingham	5-18	20958		5.4	9.9	2.1	4.2	Chest
Peckham and Butler[10]	1969	National	11	13557	4.9		12.3	2.0	3.5	General
Burr et al[11]	1973	South Wales(A)	12	817	10.2		17.0	4.2		Chest
Smith[12]	1974	Birmingham	5-6	3773		7.5		2.1		Chest
			15-16	3681		4.3		2.4		Chest
Golding and Butler[13]	1975	National	5	12977	9.8		20.8		2.1	General
Strachan and Anderson[14]	1978	Croydon(B)	7-8	4147	11.1		18.2	3.1	4.0	Chest
Lee et al[15]	1979	Newcastle	7	2700	9.3				1.2	Chest
Colver[16]	1981	Newcastle	3-11	2978			13.8	7.0		Chest
Johnston et al[17]	1984	SW London	5-13	5287	15.0		17.0		5.0	Chest
Hill et al[18]	1986	Nottingham(C)	5-11	3675	11.5			6.0	5.4	Chest
Strachan[19]	1986	Edinburgh	7	1003	12.4					Chest
Clifford et al[20]	1986	Southampton	7	1275	12.0		19.4		9.5	Chest
			11	1218	12.3		18.3		9.6	Chest
Burr et al[11]	1988	South Wales(A)	12	965	15.2		22.3	9.1	12.0	Chest
Hill et al[18]	1988	Nottingham(C)	5-11	13544	12.8			8.9	8.4	Chest
Ninan and Russell[21]	1989	Aberdeen	8-13	3403		19.8 (in last 3 years)			10.2	Chest
Symington et al[22]	1989	Walsall	7	1334	11.1					Chest
Strachan and Anderson[14]	1991	Croydon(B)	7-8	3070	12.8				7.5	Chest
Strachan et al[23]	1992	National	5-17	5472	15.0		23.0		13.2	Chest

Letters A, B, C denote studies repeated using the same methods in the same area

7.8 Greater reliance can be placed on studies where the same area was surveyed using the same methodology at two points in time. Two pairs of studies quoting annual period prevalence figures ten or more years apart have been carried out in the UK.[11,14] These are consistent with a 20-30% relative increase in the prevalence of wheezing between the 1970s and late 1980s, and a doubling in the prevalence of diagnosed asthma (studies A and B in Table 7.1).[11,14] A third pair of studies in Nottingham found a small, but statistically significant relative increase of 14% in the prevalence of wheezing in the past year between 1985 and 1988, but a larger 50% relative increase in diagnosed asthma (study C in Table 7.1).[18]

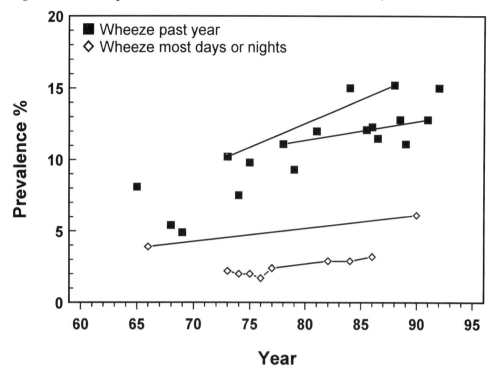

Figure 7.1 **Percent prevalence of wheeze in schoolchildren, British surveys 1965–92**

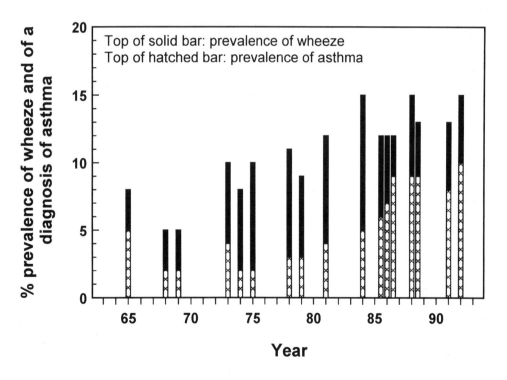

Figure 7.2 **Percent prevalence of wheeze in the past year and of a diagnosis of asthma in studies of British schoolchildren 1965–92**

7.9 Repeated studies in Aberdeen suggested a doubling in the prevalence of wheezing between 1964 and 1989,[21] but these findings are difficult to interpret due to important changes in the survey methods (the first study being based on a lengthy and multidisciplinary interview, whereas the second relied on a short postal questionnaire focusing on allergic and respiratory diseases). Furthermore, the definitions of wheezing illness used in each study were not the same, and do not permit calculation of annual period prevalence.

7.10 Each of these replicate studies was based in a small geographical area, where trends may not have reflected the national picture. A nationwide comparison has recently been made, based on parental reports of *"asthma or wheezy bronchitis in the past year"* among 16-year-olds in two national birth cohorts, born in 1958 and 1970, and studied in 1974 and 1986, respectively. The annual period prevalence obtained using this question in an older age group is much lower than most other school studies, but the relative increase, based on large sample sizes, is similar. There was a 55% relative increase (from 2.8 to 5.9%) in annual period prevalence between 1974 and 1986.[24]

7.11 Two other UK studies have reported on trends using a similar methodology at two or more time points. A different measure of disease was derived from a modification of the Medical Research Council's adult chronic bronchitis questionnaire: *"Does your child's chest sound wheezy or whistling on most days or nights?"*. Responses to this question increased (proportionately) by about 50% between the mid 1970s and mid 1980s,[25] and by a similar amount between 1966 and 1990 (see Figure 7.1).[26] However, it is possible that interpretation of this non-quantitative question may have altered over time, in line with a general increase in parental reporting of other respiratory symptoms such as cough and phlegm.[26]

7.12 The replicate surveys of 12-year-old children in South Wales are unusual in that an objective measure of bronchial hyperresponsiveness (the decline in peak expiratory flow rate after a period of exercise) was included on both occasions.[11] Reduction in peak expiratory flow rate after exercise was found to be more common in 1988 than in 1974, and the increase was particularly marked for more profound degrees of exercise-induced bronchospasm, suggesting that the severity of wheezing may have increased. However, on a formal statistical test for trend, this observation could easily have arisen by chance alone.

7.13 Although a number of cross-sectional studies of asthmatic symptoms have been carried out among British adults, comparisons between surveys conducted over the past 30 years are complicated by major changes in the content of the questionnaires used and differences in the methods of recruiting study samples. There is also uncertainty about the contribution of other diseases (such as chronic bronchitis and emphysema) to wheezing in this age group. A review of prevalence surveys of breathlessness, phlegm production and wheeze among middle aged men in Britain since the 1950s pointed out that whereas phlegm production had become less common, in line with a decline in cigarette smoking, wheezing had not.[27] Since reported wheeze was strongly related to current smoking within studies, the relative constancy of its prevalence over time would be consistent with an underlying increase in the prevalence of adult asthma, counterbalanced by a decline in chronic bronchitis and emphysema related to cigarette smoking.

Asthma: prevalence in other countries

7.14 Table 7.2 (modified from Burr[28]) lists studies of wheezing illness from other countries which have used the same or similar methods in population surveys at two points in time.[29-38] Trends in the prevalence of wheezing or diagnosed asthma have been studied among children in Switzerland,[30] the USA[33] and Taiwan.[34] Each of the three studies which assess the prevalence of symptoms suggests an increase over time. Perhaps the most convincing of these reports is the sixfold increase in prevalence among children in the rapidly developing country of Taiwan, which is based on a relevant and standardised case-definition and very large sample sizes in each survey.[34] Trends in the proportion of children diagnosed as asthmatic may reflect changes in disease labelling rather than disease activity, but there is a suggestion that the increase in prevalence of diagnosed asthma is greater among older than among younger children. A compilation of Australian prevalence studies suggests that over the past 20 years there has been an increase in the prevalence of wheeze in the past year, independent of asthma diagnosis.[29,31,32] The relative magnitude of this increase is similar to that seen in the UK, although in absolute terms the prevalence in Australian children is higher than in Britain.[39]

Table 7.2 **Serial prevalence studies of wheezing in other countries** (*Modified from Burr*[28])

Reference	Location of study	Age Group	Time period	Number of subjects	Condition	Prevalence (%)	% increase per year
Children							
Robertson et al[29]	Melbourne, Australia	7	1964	30,000	Asthma or ever	19.1	3
			1990	3325	wheeze	46.0	
Varonier et al[30]	Geneva, Switzerland	4-6	1968	4781	Recent asthma	1.7	1
			1981	3270		2.0	
Varonier et al[30]	Geneva, Switzerland	15	1968	2451	Recent asthma	1.9	4
			1981	3500		2.8	
Mitchell[31]	Lower Hutt, New Zealand	11-13	1969	952	Asthma ever	7.1	5
			1982	857		13.5	
Shaw et al[32]	New Zealand (rural maoris)	12-18	1975	715	Asthma or ever	26.2	2
			1989	435	wheeze	34.0	
Weitzman et al[33]	USA, national	0-4	1981	4083	Asthma in past	2.8	0
			1988	4959	year	2.9	
Weitzman et al[33]	USA, national	5-11	1981	5659	Asthma in past	3.6	5
			1988	6641	year	5.0	
Weitzman et al[33]	USA national	12-17	1981	5482	Asthma in past	2.8	12
			1899	6641	year	5.2	
Hsieh and Shen[34]	Taipei, Taiwan	7-15	1974	23678	Recurrent	1.3	13(a)
			1985	147373	wheezing and dyspnoea past year	5.1	
Military recruits							
Åberg[35]	Sweden, national	18	1971	55393	Asthma (denoted	1.9	4
		(males only)	1981	57150	by examining doctor)	2.8	
Haahtela et al[36]	Finland, national	19	1966 to	c. 38000	Asthma (denoted	0.3	8(b)
		(males only)	1989	annually	by examining doctor)	1.8	
Laor et al[37]	Israel, national	17-18	Not	134863	Asthma (denoted	1.7	20
		(male,	stated	144491	by examining	2.2	
		female)	(c)	163832	doctor)	3.3	
Adults							
Peat et al[38]	Bussleton, W Australia	18-55	1981	553	Wheeze ever	26.4	4 (d)
		(male,					
		female)	1990	1028		36.7	

(a) Recurrent defined as more than three episodes in the year
(b) In 1961 the prevalence was estimated at 0.08%, similar to that in the 1920s and 1930s
(c) Three consecutive 3-year periods. Exact dates not given for reasons of national security
(d) Bronchial hyperresponsiveness to inhaled histamine was slightly *less* common in 1990

7.15 Three studies of military recruits in Scandinavia[35,36] and Israel[37] provide information on trends in asthma as diagnosed by a standard medical examination in large, fairly representative national samples of young adults. Each shows evidence of an increase over time, but the 20-fold increase in asthma recorded among Finnish army recruits over a 30-year period is the most remarkable. This change is too large to be plausibly explained by changes in symptom reporting or disease labelling.[36]

7.16 The only study of adults at two points in time is from Busselton, Western Australia.[38] Among men and women aged 18-55 years there was a relative increase of 39% in the lifetime prevalence of wheeze and an 81% increase in diagnosed asthma between 1981 and 1990. Unfortunately, the question used to define recent wheeze was not strictly comparable between the two surveys. In contrast, there was a non-significant *decrease* in the proportion of subjects with bronchial hyperresponsiveness assessed by inhaled histamine challenge. Thus, the possibility arises that the increase in prevalence of symptoms was a reflection of a greater public awareness of wheezing illness following national health education campaigns highlighting asthma, rather than a change in the underlying prevalence of disease.

Asthma: severity

7.17 Few of the serial prevalence studies have obtained comparable information on disease severity. Two measures were available in the nationwide comparison of 16-year-olds from the British 1958 and 1970 birth cohorts.[24] The prevalence of frequent attacks of asthma or wheezy bronchitis (defined as attacks occurring more than once a week) more than trebled, from 0.2% in 1974 to 0.7% in 1986. This was a significantly greater proportionate increase than for annual period prevalence. However, the proportion of children who had been absent from school due to wheezing illness over the past year was little changed (1.5% in 1974 compared to 1.7% in 1986).

7.18 The most extensive information on time trends in asthma severity is that reported from replicate surveys of 8-year-old children in Croydon, south London in 1978 and 1991.[14,40] Here, the prevalence of frequent attacks of wheezing was almost the same in each survey,[14] and other methods of morbidity did not show a consistent trend in either direction.[40] The prevalence of sleep disturbance due to wheezing, restriction of activities at home and the prevalence of speech limiting attacks of wheeze had almost doubled, whereas prolonged school absence due to wheezing, restriction of activities at home and the prevalence of speech limiting attacks of wheeze (characteristic of episodes which might be considered deserving of hospital admission) fell by one-half. Although this latter change was not quite statistically significant, it argues against a marked increase in the occurrence of intense attacks of wheezing.

Asthma: use of health services

7.19 Trends in hospital admissions and general practice contacts are more difficult to interpret as they reflect not only trends in morbidity (which may be influenced by epidemiological forces and/or by changes in treatment),[41] but also a range of other factors, including shifting patterns of presentation, diagnosis and referral, and changing preferences for inpatient or domiciliary care. It is also important to distinguish between measures based on the number of *persons* using the system and those related to the number of *episodes* of care which are provided. An increase in episode-based measures may be due to more frequent consultation or admission of the same number of patients.

7.20 A rising rate of hospital admissions for childhood asthma in England and Wales from 1958 onwards provided one of the earliest signs that the disease might be on the increase.[42,43] From 1958 to 1985, national trends were based on a random 10% sample of discharges and deaths in which diagnostic coding was known to be virtually complete (Hospital Inpatient Enquiry—HIPE—see Tables 7A.3 and 7A.5 in Annex 7A). Subsequently, there has been a transition to 100% coverage of hospital activity (Hospital Episode System: HES), but diagnostic coverage was incomplete in the early years and this complicates the interpretation of more recent trends, which are discussed in detail later.

7.21 Asthma admission rates increased 13-fold in the 0-4 age group and six-fold in the 5-14 age group between the early 1960s and the mid 1980s (see Figure 7.3). This rise does not appear to be explained by an increase in readmissions or a substantial change in the clinical severity of disease which prompted admission.[44] Only a small part of the increase can be accounted for by changes in diagnostic labelling, as rates of admission for alternative disease categories such as acute or unspecified bronchitis, or pneumonia, have been fairly stable and are substantially lower than the asthma admission rate.[42,44] An increasing proportion of all paediatric admissions have been due to asthma.[45]

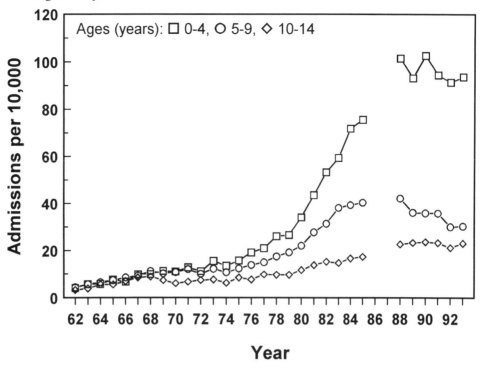

Figure 7.3 **Hospital admission rates for asthma, 0-14, both sexes, England & Wales 1962–85, and England only 1988–93**

7.22   There is evidence of changes in the pattern of medical care for acute asthma in Britain in recent decades. As inpatient care for other diseases has become less common, supply of paediatric beds has improved. Several studies suggest an increased demand for admission through self-referral to accident and emergency departments during acute episodes of asthma in children.[14,44,46] These changes have occurred against a background of fewer home visits by general practitioners[14] and an increasing tendency for GPs to refer acute asthma to hospital.[47] All of these forces may have tended to increase the likelihood that acute episodes of wheezing result in admission to hospital, independent of any change in the occurrence of intense attacks.

7.23   Trends in hospital admission rates in adults are less marked and their direction has varied over time, declining slightly in the early 1970s before resuming an upward rise in the 1980s (see Figures 7.4 and 7.5). The possible influences of changing diagnostic preference, referral patterns and conventions for clinical management have not been assessed among adults. As discussed above, the underlying trends in prevalence are also unclear in this age group.

7.24   Interpretation of trends in admission rates in England throughout the 1980s is severely complicated by the transition from HIPE to HES, and uncertainties about the completeness and validity of early HES data. No in-patient data is available for 1986 and early 1987 in England, but adjustment factors have recently been published which permit an estimate of the number of total admissions in 1988–91. These adjustments must be considered imperfect as the grossing up factor has been derived for all age groups and diagnostic categories, whereas incompleteness of diagnostic coding may well have varied by specialty and client group. The adjusted trend data for England (only) are presented in Tables 7A.4 and 7A.5 of Annex 7A. There appears to have been little change in adjusted admission rates in most age groups over this period.

7.25   A more rigorous assessment of trends in admission rates through the 1980s is provided by published data for Scotland and Wales. In both countries, 100% coding of inpatient diagnoses preceded the demise of HIPE and continued until 1990 in Wales and still continues in Scotland, where the most recent published data related to

1992. In Scotland, asthma admission rates in children and young adults started to rise in the mid to late 1970s, somewhat later than in England and Wales.[48] The upward trend in admission rates in all age groups continued through the 1980s in both Scotland and Wales, although in the most recent data for each country there is a suggestion that admission rates may have levelled off (Scotland), or started to decline (Wales) (see Table 7A.4 of Annex 7A). The apparent plateau in English rates would be consistent with these patterns. A special investigation of asthma admission rates in East Anglia suggests that a decline commenced in that region in the late 1980s.[49] The reasons for this reversal, if it proves to be sustained, are poorly understood.

Figure 7.4 **Hospital admission rates for asthma, ages 15-44, both sexes, England & Wales 1962–85, and England only 1988–93**

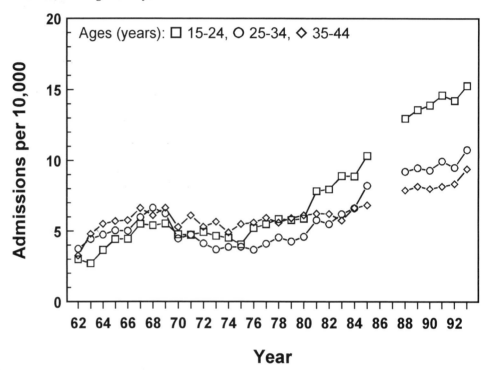

Figure 7.5 **Hospital admission rates for asthma, ages 45+, both sexes, England & Wales 1962–85, and England only 1988–93**

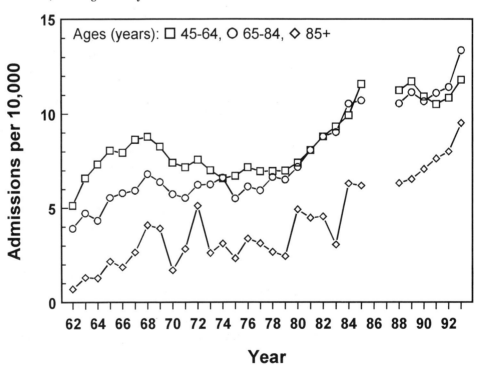

7.26 Two sources of data are available relating to general practice contacts for asthma. The number of persons consulting in one year can be derived from four national morbidity surveys which were conducted in 1955–56, 1970–72, 1981–82 and 1991–92. These suggest a marked increase in the proportion of the population in all age groups consulting for asthma (see Figure 7.6 and Table 7A.2 in Annex 7A).

Figure 7.6 **Patients consulting their GP for asthma by sex and age, England and Wales 1971–2, 1981–2, 1991–2**

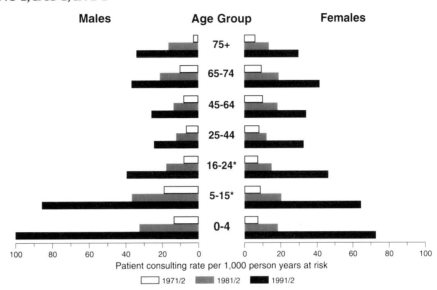

* For 1971/2 and 1981/2, age groups are 5-14 and 15-24.

Analysis of the second and third surveys suggested that this could not be entirely explained by changes in the catchment population, nor by diagnostic transfer from alternative disease categories, such as acute or chronic bronchitis.[50]

7.27 The second source is a continuous series of weekly records kept by volunteer "spotter" practices participating in the Royal College of General Practitioners' sentinel practice scheme. Practitioners are asked to record consultations for selected diagnoses, including "acute asthmatic attack", "acute bronchitis" and "hay fever", so the measure of disease occurrence (the mean weekly attack rate) is an episode-based, not a patient-based measure. The panel of practices participating in the scheme have changed from time to time, which complicates the assessment of time trends. Nevertheless, there has been a striking increase in the mean weekly attack rate for acute asthma in all age groups from 1976 to 1992 (see Figure 7.7). The increase is particularly marked in the under-fives (more than a tenfold increase) and 5-14 age group (more than a fivefold increase). Trends for acute bronchitis have also been upward,[51] suggesting that diagnostic transfer is unlikely to explain the recorded increase in asthma attack rates.

7.28 Little is known of changes in the consultation behaviour of patients with asthma over the past 20-30 years. A high proportion of wheezy children had been treated by their general practitioner at the surgery in both 1978 (77%) and 1991 (84%),[14] but the tendency to consult in response to acute symptoms has not been studied specifically. The limited evidence from prevalence studies does not suggest a marked increase in the frequency or severity of asthmatic attacks, at least among children, so the increase in health service activity in both primary and secondary care may well reflect changing patterns of service utilisation rather than an increase in the incidence of troublesome wheezing episodes. This interpretation, however, should not be taken to imply dismissal of the evident rise in *workload* related to asthma in both hospital and general practice in Britain over the past three decades.

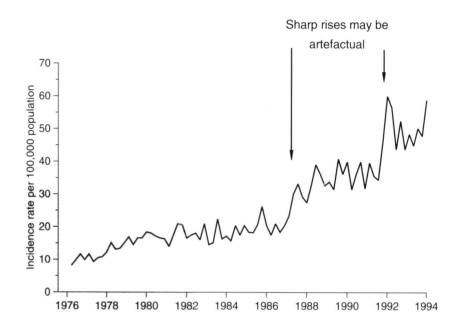

Figure 7.7 **Trends in asthma reported by GPs. Mean weekly incidence in 12 week periods. All persons, all ages, England and Wales, 1976–1993**

7.29   An upward trend in health care utilisation for asthma has been noted in other countries, including the USA,[52] Canada,[53] Australia and New Zealand.[54] The reasons for these trends in other countries have not been as thoroughly evaluated as in the UK.

Asthma: mortality

7.30   Death from asthma is fortunately a rare event. Recently, about 2,000 deaths each year have been certified as due to asthma, more than half of them among adults aged 65 or over. While trends in asthma mortality rates have attracted a lot of attention, they are influenced by many non-biological factors, which makes them a poor guide to underlying changes in the prevalence or severity of the disease. These other factors include changes in: the rules for certifying and coding deaths; diagnostic preferences; patient management, including drug treatment; patient behaviour, including compliance with medication and awareness of the need to seek help in the event of an acute attack of asthma.

7.31   Mortality rates for asthma in England and Wales have fluctuated over the past 35 years (see Figure 7.8). During the 1960s there was an "epidemic" of asthma deaths which affected young children and young adults most in relative terms. This rise was thought to be related to the introduction of a new aerosol preparation (high-dose isoprenaline) and subsided after the Committee on Safety of Medicines issued a warning on its use in 1966.[55] It remains controversial whether the rise in mortality was a direct result of drug toxicity, or due to delay in presentation by patients placing unwarranted faith in its clinical effectiveness.

7.32   The asthma mortality rate fell during the early 1970s, but began to rise again through the 1980s (see Figure 7.8). Part of this increase appears to have been related to changes in the method of coding deaths (a change in International Classification of Diseases in 1979 and introduction of a new convention for coding deaths due to pneumonia in 1984). Mortality rates among the elderly were particularly affected, and it is likely that there was greater preference for asthma as a diagnostic label for chronic obstructive airways disease in this group. Certainly, death rates from asthma are low by comparison with those from chronic bronchitis, emphysema and obstructive airways disease, and only a small diagnostic transfer from these categories into asthma could have accounted for the rise in asthma mortality rates (see Figure 7.9).

Figure 7.8 **Age specific asthma mortality rates, both sexes combined, England & Wales 1958–92**

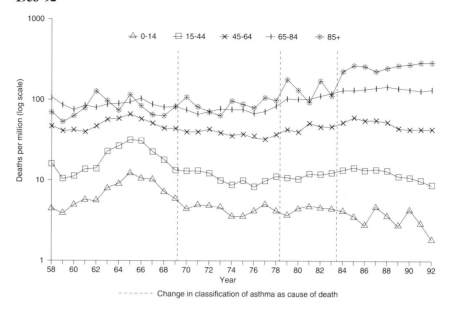

Figure 7.9 **Chronic obstructive pulmonary disease mortality by cause, age 65–84, England & Wales 1971–92**

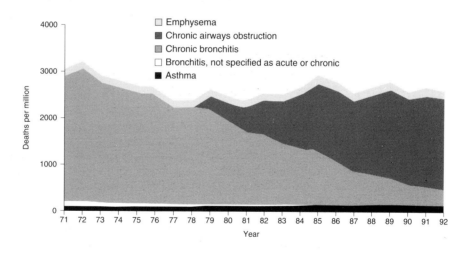

7.33   Over the past three years, mortality rates for asthma have declined in all age groups. The reasons for this reversal of the earlier trend are unclear (see Figure 7.8). Nevertheless, despite many therapeutic advances, current asthma mortality rates among children and young adults (the group in whom alternative diagnoses are least common and management ought to be most effective) are at a similar level to those recorded in England and Wales in the 1930s.[43,56]

Asthma: overview

7.34   Table 7.3 summarises the trends in asthma prevalence, health service use and mortality in pre-school children, school children and adults in Britain over the past 30 years.

Table 7.3 **Summary of time trends in asthma in the UK 1962–92**

	Age Group		
	0-4	**5-14**	**Adults**
Prevalence	No data#	× 1.5	No data
GP Contacts	× 10	× 5	× 3
Admissions	× 20	× 10	× 2-3
Mortality	Low and falling		Variable

#Indirect evidence from lifetime prevalance in school suggests no dramatic increase (see Table 7.1)

The most rapidly increasing burden on the health service is among pre-school children, for whom there is a regrettable lack of direct prevalence data. However, data obtained from school surveys do not suggest a marked increase in the lifetime prevalence of wheezing (see Table 7.1), so it is likely that the increase in health service contacts for asthma among pre-school children is a substantial exaggeration of any upward trend in prevalence.

Figure 7.10 **Time trends in indicators of asthma among British schoolchildren, 1962–92**

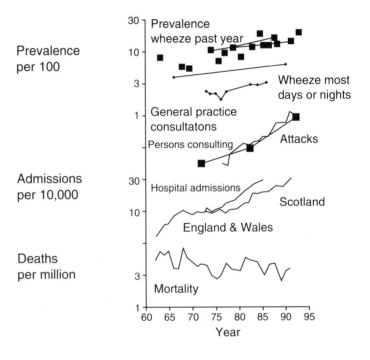

7.35 The data are most extensive for school children. Available time series for prevalence, general practitioner consultations, hospital admissions and mortality among 5-14 year old children in the UK over the past 30 years are collated in Figure 7.10. A logarithmic scale is used to compare proportionate changes in each measure of disease burden. It is clear that the relative increase in use of health services is much greater than any rise in prevalence. This discrepancy may reflect a change in the epidemiology of asthma, with a proportionately greater increase in the more severe forms of wheezing, although the evidence on this point is inconsistent. Alternatively, it may reflect the influence of changes in medical care of acute asthmatic episodes.

7.36 Taken overall, the evidence is in favour of an increase, rather than a decrease, in the prevalence of wheezing illness in the UK. However, among school children, where the data are most extensive, the magnitude of the increase is modest by comparison with the upward trend in hospital admissions and primary care contacts, and there is no consistent evidence of a change in the severity of disease. The upward trend in the proportion of children diagnosed with asthma during the 1980s almost certainly overestimates any change in disease prevalence in this age group, and is probably attributable to changing professional and lay preference for asthma as a diagnostic label. It is probable that both epidemiological and medical care factors have contributed to the increasing burden on health services in this age group.

7.37 The results of the more recent school surveys suggest that the increase in annual period prevalence of wheezing may have levelled off (at about 11-15%), and hospital admission rates for asthma in England, Scotland and Wales may also have reached a plateau in the last few years. Mortality rates have fluctuated over recent decades but are currently falling. The direction and gradient of future trends in asthma in the UK are, therefore, uncertain.

7.38 Future trends in asthma might be predictable if the propensity to wheeze was largely determined in the early years of life. The occurrence of asthma in each generation could then be predicted to some extent throughout life. Trends in asthma mortality[56] and hospital admissions[57] have been interpreted as evidence of such a cohort (or generation) effect. However, closer examination of age-specific admission rates show that the predominant influence is related to the year of admission, not year of birth, with fluctuations occurring simultaneously in a wide range of age groups (Figures 7.3-7.5). Similarly, the major variations in asthma mortality rates are those related to period of death (particularly the 1960s "epidemic"), rather than period of birth (Figure 7.8). It is, therefore, unlikely that the future burden of asthma will be predictable to any great extent by extrapolation from the past, although it is possible that cohort influences are important in determining upward trends in allergic diseases in general. These are discussed in the next section.

Other allergic diseases

7.39 There are few series of comparable data on the prevalence of allergic diseases other than asthma over time, and these are subject to similar biases as those for wheezing illness, due to changes in awareness and labelling of symptoms, by both the public and the medical profession.

7.40 There is anecdotal evidence that hay fever was exceedingly rare in Britain before the industrial revolution.[58] Dr John Bostock, who described his "periodical affection of the eyes and chest" in the medical press in 1819, took another nine years to find 27 other cases of what had, by then, become known as *"cattarhus aestivus"* (summer catarrh).

7.41 A large population survey in Switerzland in 1926 reported that only 0.3% of 77,000 subjects of all ages suffered from hay fever.[59] The prevalence in Zürich at that time was 1.2%, rising to 4.8% in 1958 and 10% in 1985.[59] Each of these studies was based on a large sample of all ages, although the definition of hay fever differed somewhat between surveys.

7.42 Compilation of a series of surveys of American college students over the period 1924 to 1969 also suggests an increase in prevalence of hay fever throughout this century.[60] The combined lifetime prevalence of asthma, hay fever and allergic rhinitis rose from 3.3% in 1924 to 25% in 1969, apparently mainly due to an increase in the prevalence of hay fever or allergic rhinitis. Some of these surveys were of small samples and they were performed in different institutions without a standardised case definition.

7.43 Perhaps the best data supporting an increase in prevalence of allergic rhinitis more recently relate to medical examinations of 18-year-old Swedish army recruits performed in 1971 and 1981.[35] A current tendency to allergic rhinitis was diagnosed in 4.4% of 55,393 conscripts in 1971 and 8.4% of 57,150 conscripts in 1981. Further smaller studies of school children in two areas of Sweden suggest a continuing rise in hay fever prevalence throughout the 1980s.[61]

7.44 The proportion of the British population consulting their doctor for hay fever rose from 5 per 1,000 in 1955–56, to 11 per 1,000 in 1970–71 and 20 per 1,000 in 1981–82,[50] but this may reflect more widespread recognition and labelling of symptoms, or the development of effective treatments available on prescription. Throughout the 1980s there was little consistent trend, although the summer of 1992 was an unusually high year for consultations related to hay fever.[62]

7.45   Three studies of British teenagers suggest that the prevalence of hay fever, as reported by parents, has increased substantially since the 1960s. In Aberdeen, the prevalence among 10-14 year olds was 3% in 1964 and 12% in 1989.[21] Although identical questions were used, there were important differences in the survey techniques in these two studies, as discussed previously, which may have led to more effective ascertainment of allergic diseases in the 1989 survey. However, comparison of the 1958 and 1970 national birth cohorts found a similar increase in prevalence of hay fever in the past year at age 16: from 12% in 1974 to 23% in 1986, each based on questions embedded in a lengthy interview with parents.[24] In the smaller, but more focused, studies of South Wales children, the lifetime prevalence of hay fever among 12 year olds was 9% in 1973 and 15% in 1988.[11]

7.46   Eczema in early childhood is often allergic in nature. The lifetime prevalence of eczema, as recalled by parents when their children were aged 6, 7 or 5, respectively, was compared in the British 1946, 1958 and 1970 birth cohorts.[63] The prevalence rose from 5.1% in the 1946 cohort to 7.3% in the 1958 cohort to 12.2% in the 1970 cohort. The prevalence of eczema in the past year at age 16 in the 1958 and 1970 cohorts show a similar increase, from 3.1% to 6.4%.[24] In Aberdeen, the lifetime prevalence of eczema in 10-14 year-olds more than doubled from 5% in 1964 to 12% in 1989,[21] and among 12 year olds in South Wales, it trebled from 5% to 16%.[11] A study of Danish twins has also suggested an increase in prevalence of childhood eczema over recent years.[64]

7.47   Studies using skin prick tests of aeroallergen sensitivity provide objective evidence concerning trends in allergic sensitisation. Over 1,300 residents of Tucson, Arizona were tested by skin prick to common aeroallergens at a mean interval of eight years, over which time the prevalence of cutaneous reactivity increased from 39% to 50%. This increase was evident in all age groups,[65] whereas in each survey analysed cross-sectionally the prevalence of skin prick positivity declined with increasing age from middle age onwards. This suggests that the low prevalence in the elderly may reflect a cohort (or generation) effect: earlier generations carrying a lower propensity to allergy throughout their life, rather than a biological effect of ageing. The Tucson findings are consistent with results from south London, where two cross-sectional surveys using similar methods in adjacent (but not identical) areas found that the prevalence of positive skin prick responses rose significantly from 23% in 1974 to 46% in 1988.[66] In contrast, there was little change in the proportion of adults in Busselton, Western Australia who reacted on skin prick testing to local allergens: 39% in 1981 compared to 41% in 1990.[38]

7.48   The results of skin prick tests are prone to spurious variations due to differences in operator technique and allergen potency which are difficult to standardise over long periods of time. These methodological difficulties were overcome by a Japanese study which measured allergen-specific IgE in serum samples stored from population surveys of schoolgirls in 1978, 1981, 1985 and 1991. The prevalence of allergic sensitisation to one of fourteen common aeroallergens and two food allergens increased significantly from 21% in 1978 to 39% in 1991.[67]

7.49   Although none of this evidence is particularly strong in isolation, when considered together it suggests that if asthma is becoming more common, it may be related to a rise in the prevalence of atopy in general, rather than to changes specific to the lung.

Air pollution

7.50   There have been major changes in the nature of air pollution in the UK over the past 40 years; urban smoke and SO_2 (winter smog) have declined and primary and secondary pollutants related to vehicle emissions may be increasing. However, as discussed in Chapter 2, measured levels of vehicle-related pollutants at sites with records dating back ten years or more, do not suggest an increase in ambient exposure to primary (NO_x) or secondary (ozone) pollutants.

7.51   As a result of the Clean Air Act passed in 1956, urban smoke and SO_2 concentrations declined throughout Britain from the early 1960s, and continued to fall throughout the 1980s (see Figure 7.11).

Figure 7.11   **Smoke and sulphur dioxide: trends in urban concentrations**

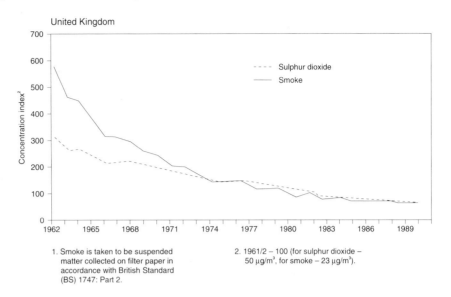

United Kingdom

1. Smoke is taken to be suspended
   matter collected on filter paper in
   accordance with British Standard
   (BS) 1747: Part 2.

2. 1961/2 – 100 (for sulphur dioxide –
   50 µg/m³, for smoke – 23 µg/m³).

Earlier data are patchy, because it was not until 1962 that the national network of filters and peroxide samplers for daily measurement of smoke and SO_2 levels was widely established. However, results for London, where Clean Air legislation probably took effect earlier than in other conurbations, suggests little change in annual mean SO_2 concentrations from the 1930s to the late 1950s. (see Fig. 7.12)[68]

Figure 7.12   **Annual mean sulphur dioxide (SO_2) concentrations measured at County Hall, London (1931–1985)**

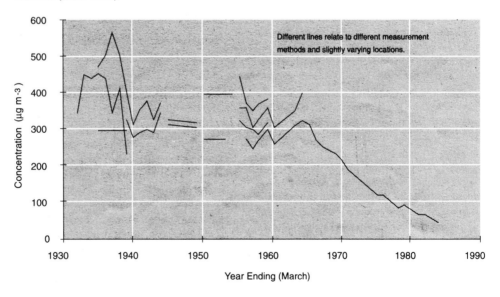

Different lines relate to different measurement methods and slightly varying locations.

Year Ending (March)

7.52   The effect of the decline in urban concentrations has been to remove much of the urban-rural difference in SO_2 concentrations (see Figure 7.13), although particulate pollution remains higher in urban areas, where diesel exhausts (rather than domestic or industrial sources) now contribute the majority of airborne particles.[69]

Figure 7.13 **Comparison of SO₂ concentrations measured in the City of Lincoln and rural Lincolnshire**

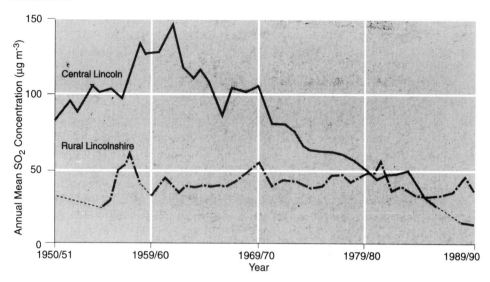

7.53 Evidence from three sites which have recorded ozone levels since the 1970s (Sibton, Stevenage and Central London) is consistent in showing a *downward* trend in annual mean, annual 98th percentile and number of hours exceeding 60 ppb (see Figure 7.14).[70]

7.54 There has been an increase in emissions of NO_x from vehicular sources and some evidence of an increase in peak concentrations of NO_2. There is little evidence of a substantial change in annual mean urban concentrations measured in central London (arguably traffic saturated since the mid-1970s), Glasgow or Manchester (see Figure 7.15).[68] On the other hand, national surveys of NO_2 carried out using passive samplers in 1986 and 1991 suggested somewhat higher levels of exposure throughout the country on the second occasion.[71,72]

7.55 Personal exposure to specific pollutants depends on many factors other than ambient levels, including indoor sources of pollutants (particularly NO_x from unvented gas appliances and paraffin heaters and particles from open fires and tobacco smoking), ventilation of buildings, duration of time spent outdoors, "kerbside" exposure and the duration and intensity of outdoor exercise. Little is known of changes in these factors over time.

Comment

7.56 There is little evidence to support a link between long-term time trends in asthma and changes in air pollution exposure in the UK. Trends in the major air pollutants vary dependent on source. In urban areas ambient levels of NO_2 are rising or stable. Levels of Black Smoke have declined dramatically over the last 30 years, although levels now seem stable. Long term average levels of ozone have decreased in urban areas, but at more rural sites have increased. The best available evidence on trends in asthma suggests an upward trend during the 1970s which may have levelled off in the 1980s. Trends in health service utilisation and diagnosed asthma are likely to have been affected more by changes in disease labelling, consultation and referral patterns than by changes in the prevalence or severity of wheezing illness in the community.

7.57 It is plausible to argue that there has been a general increase in allergic disease in industrialised countries in the postwar period, leading to a rise in the prevalence of atopic asthma. It is possible that this has been balanced, to a varying extent in different age-groups, by a decline in other forms of wheezing.

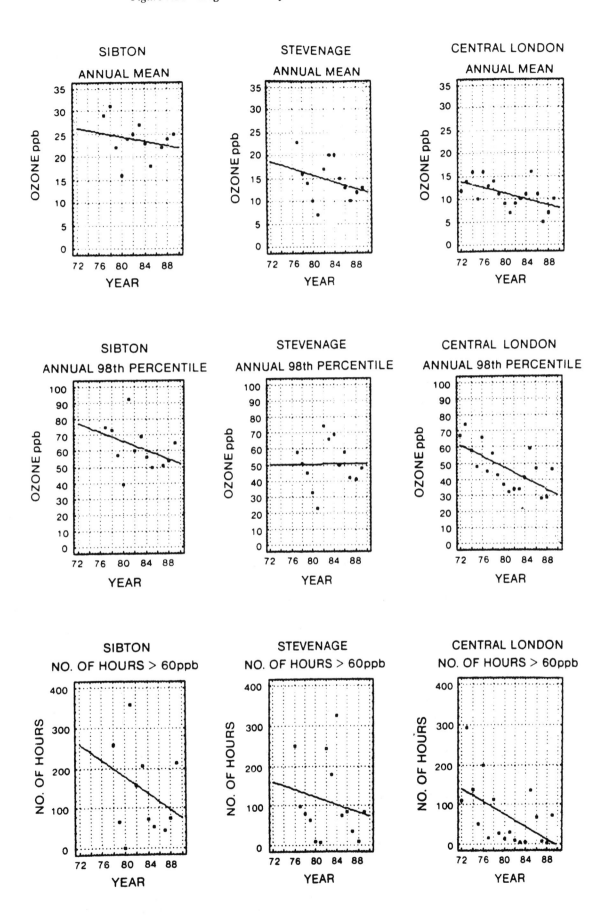

Figure 7.14 **Regression analysis of ozone time series, 1972–1989**

7.58   Although smoke and SO_2 levels have declined dramatically over the past 30 years, the generations who are young adults in the 1990s may still have been exposed to "old-fashioned" air pollution in the first few years of life, particularly if they were born in urban areas. Possible long-term effects of this early exposure may influence regional and urban-rural differences in the prevalence of wheezing illness which are discussed in later sections.

Figure 7.15   **Annual mean concentrations of NO and NO_2 in London (Victoria), Glasgow, Manchester and London (County Hall rooftop) in recent years, for which data are available**

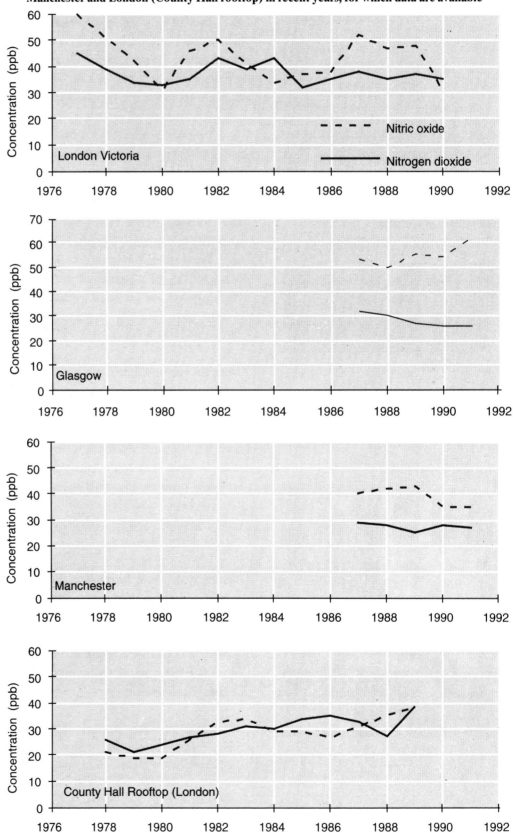

Asthma and related diseases

7.59 The rate of occurrence (incidence) of asthma attacks in Britain varies considerably throughout the year, but the seasonal pattern differs according to the age group studied. Different patterns are observed for health service contacts and mortality.

Figure 7.16 **Average 4–weekly percentage variation from the trend in hospital admissions for asthma and new GP episodes of asthma by age**

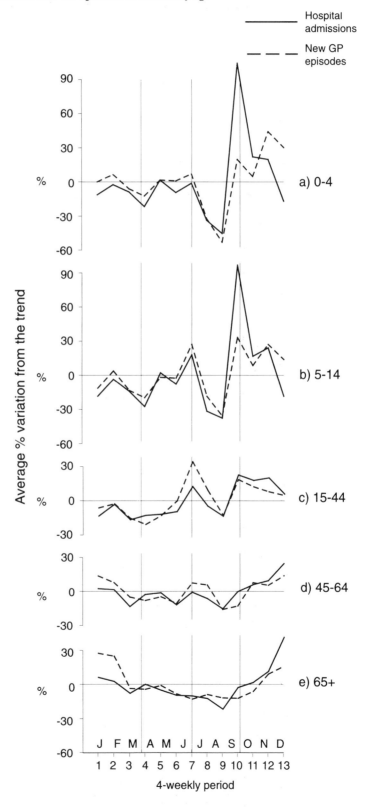

7.60   "Acute asthmatic attack" is one of several diagnoses routinely reported by a sample of "spotter" practices participating in the Royal College of General Practitioners Weekly Returns Service. This network provides a series of weekly incidence rates relating to a population of about 200,000 people. Asthma attack rates in all ages show a peak in late June-early July, some 2-3 weeks after the peak concentrations of airborne grass pollen, with a second peak in early autumn.[73] The seasonal pattern for acute bronchitis is different, with a marked winter peak, and absence of a rise following the pollen season.[74] Consultations for hay fever, as might be expected, mainly coincide with the grass pollen season.

7.61   Hospital admissions for asthma and for acute bronchitis show similar seasonal patterns to those for general practice contacts for each diagnosis.[75] Figure 7.16 compares the seasonal pattern for asthma in different age groups in recent years, as determined from the Weekly Returns Service and Hospital Episode System. The cyclical pattern differs substantially by age, with a marked September peak in children, a lesser rise in young adults, and no autumn peak in older adults. The early summer (pollen season) peak is evident only in school children and young adults. A winter peak in asthma admissions is apparent in older adults, but this may be exaggerated in these data by a peak of asthma admissions in the over-45s associated with the 1989 influenza epidemic (see below).

7.62   In contrast to these seasonal patterns, death from asthma in the 5-34 age-group is least common in winter and peaks in August, when the hospital admission and general practice consultation rates are at one of their lowest points.[76]

7.63   Figure 7.17 shows the variation in weekly admissions for asthma in England over four years (April 1987 to February 1991).

Figure 7.17 **Weekly number of hospital admissions for asthma by age, England 1987/88–90/91**

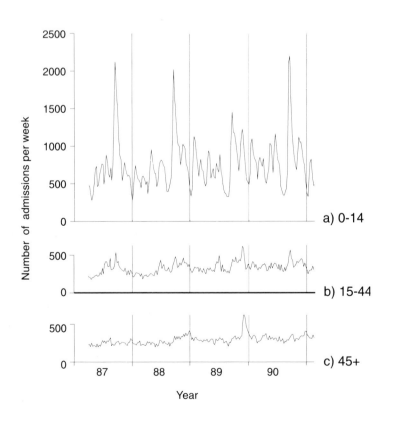

The periodicity of admissions in children is very marked and fairly regular from year to year. This complex cyclical variation is closely related to school holidays, during which admissions and general practice consultations decline. There is also evidence of a decline in admissions during school half-term holidays.[77] This phenomenon is unlikely to be due entirely to a reduction in the population at risk, related to holidays abroad, since children would be expected to take holidays with their parents, and the relative decline during holiday periods is much less marked in young adults. It has generally been attributed to the interruption of virus transmission in schools during holiday periods.[77] The role of virus infections in precipitating attacks of wheezing in both children[78,79] and adults[80] is well established, and epidemics of viral infections are probably more important than naturally occurring peaks of aeroallergen exposure in determining the temporal pattern of wheezing episodes in children.[81]

7.64   Figure 7.17 also demonstrates the marked effect of the 1989 influenza A epidemic on asthma admissions in adults. This epidemic occurred during the last seven weeks of 1989 and was associated with an excess in general practitioner consultations for all forms of respiratory disease.[82] The hospital data show a doubling in admission rates at the peak of the epidemic among adults over 45 years of age. A less clearly defined peak of comparable magnitude occurred among younger adults.

7.65   An autumn peak in hospital admissions for asthma has been observed in other countries. In New York, well-defined peaks of admissions occurring in September of several years from 1957 to 1965 were associated with the onset of colder weather and the first requirement for domestic heating since the spring.[83,84] It was speculated that cold, dry air might exacerbate existing asthma. However, alternative explanations could be that interpersonal contacts, facilitating virus transmission, were more common in colder weather when families clustered indoors, or that the lower humidity of heated indoor air influenced the probability of virus transmission.[85-88]

7.66   In general, influences of weather on the occurrence of asthma have been largely attributed to the effect of climatic conditions on aeroallergen exposure.[89] However, in the isolated island of Bermuda, where aeroallergen exposure is extremely low, emergency room attendances with acute asthma were more common on cold, less humid days, typically associated with a northeasterly wind. Since this wind blows from the mass of the Atlantic Ocean, this was considered evidence of a direct effect of climatic conditions on asthma incidence.[90]

Air pollution

7.67   Historically, air pollution levels fluctuated markedly in British towns, due to higher emissions of smoke and SO_2 in winter months from domestic coal fires. As low level emissions from domestic sources were controlled by Clean Air legislation, this seasonality diminished greatly[91] and in recent years, SO_2 levels have been similar in winter and summer, with peaks occurring at any time of year (see Figure 7.18). Where coal-fired heating remains in use, as in Belfast, the older pattern of winter peaks persists.

7.68   The seasonal variation in NO_2 levels is not marked, but peak concentrations are more likely to occur in winter. The time series from central London in recent years is dominated by the unprecedented levels recorded during a pollution episode in December 1991 (see Figure 7.18), which is discussed in greater detail later.

7.69   Ozone levels display a marked seasonality, with higher levels in the summer months, related to more prolonged and intense sunlight which catalyses the formation of this secondary pollutant. Peaks of ozone pollution are also more common in the summer months (see Figure 7.18).

Comment

7.70   The seasonal pattern of asthma attacks is complex and bears little relationship to cyclical variations in the major pollutants. The early summer peak of asthma incidence precedes the period of maximum ozone levels, and the early autumn asthma peak occurs after ozone concentrations have begun to fall. Alternative explanations are available to explain the periodicity of asthma and it seems unlikely that air pollution exposure is an important determinant of these regular patterns.

Figure 7.18   **Daily maximum hourly levels of NO₂, SO₂ and ozone in central London**

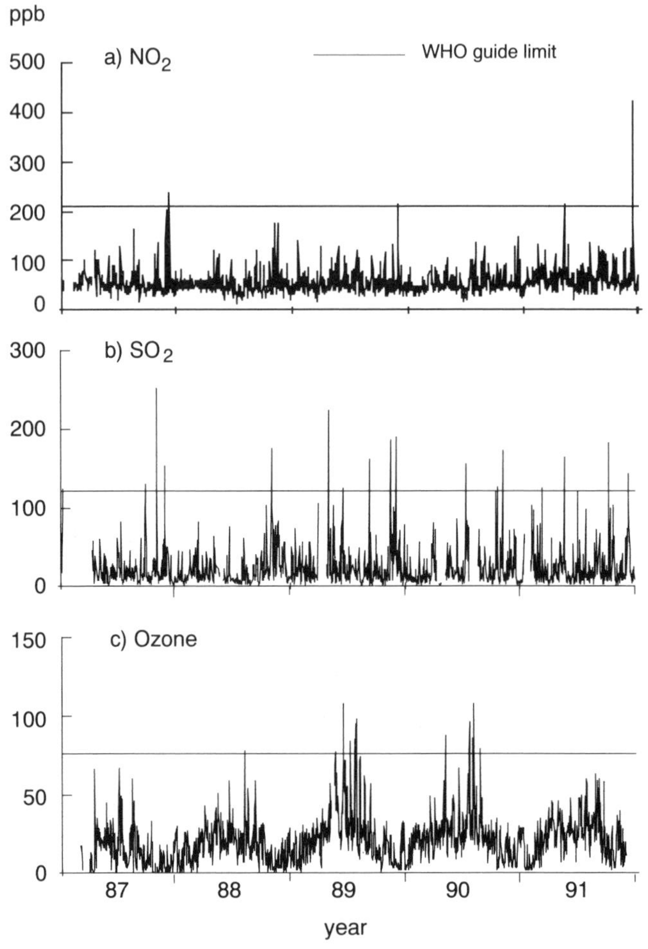

Air pollution

7.71   There is currently no comprehensive network of air pollution monitoring sites throughout Britain. However, the regional pattern of exposure to certain pollutants may be estimated or modelled from the data available from existing monitoring sites, combined with knowledge of atmospheric chemistry and prevailing meteorological conditions. In the case of NO_x, data from a nationwide survey using passive diffusion tube samplers was used to predict the pattern of exposure on the basis of local population density.[72]

7.72   Figures 7.19a to 7.19c and Table 7A.8 (in Annex 7A) illustrate the estimated distribution of air pollutants in Britain in the early 1990s.

Each shows a characteristic pattern: SO_2 levels are highest around the Wash, reflecting high-level emissions by power generation in the Trent valley and a prevailing westerly wind; ozone levels show a north-south gradient, mainly reflecting duration and intensity of sunlight; NO_x are concentrated in urban areas, reflecting their emission as primary pollutants from motor vehicles.

7.73   The current pattern of pollution differs greatly from that in the past, when smoke and SO_2 levels were generally higher than today, and particularly high in urban areas in winter, reflecting low-level emissions from domestic coal heating. During the winter of 1962–63, the first season when the national network of daily

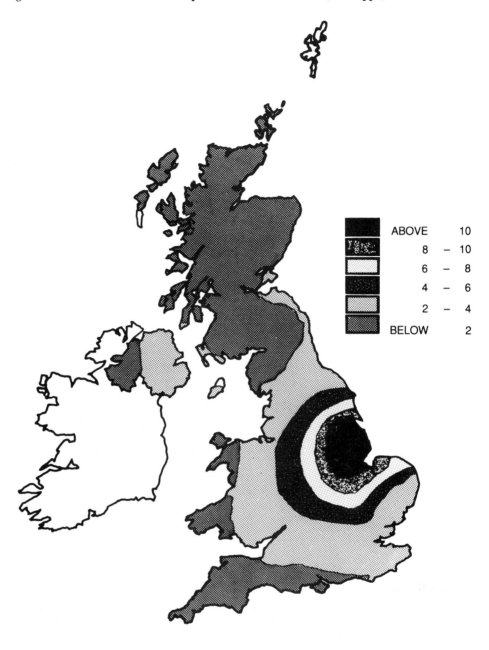

Figure 7.19a   **Annual mean rural sulphur dioxide concentration, 1990 (ppb)**

■	ABOVE	10
▨	8 –	10
▫	6 –	8
▦	4 –	6
░	2 –	4
▓	BELOW	2

smoke and SO_2 monitors was well established, six-monthly mean levels of smoke and SO_2 levels were above 200 µg/m³ in many major towns and cities and daily average concentrations above 1000 µg/m³ were not uncommon (see Table 7A.7 in Annex 7A).

Asthma

7.74   The data which are available on regional variations in asthma prevalence relate mainly to school children. Annual period prevalence data from the British 1958 and 1970 birth cohorts[92] suggests that the prevalence of wheezing at age 5-7 years was lowest in Scotland, and highest in southwest England and Wales (see Figure 7.20). The range of prevalences was approximately twofold, but some of this may have been due to chance fluctuations with a modest sample size in each region. Regional differences became less apparent with advancing age in the 1958 cohort, and there was little evidence of regional variation in the prevalence of asthma or

Figure 7.19b   **Estimated NO₂ concentrations (ppb) in Great Britain**

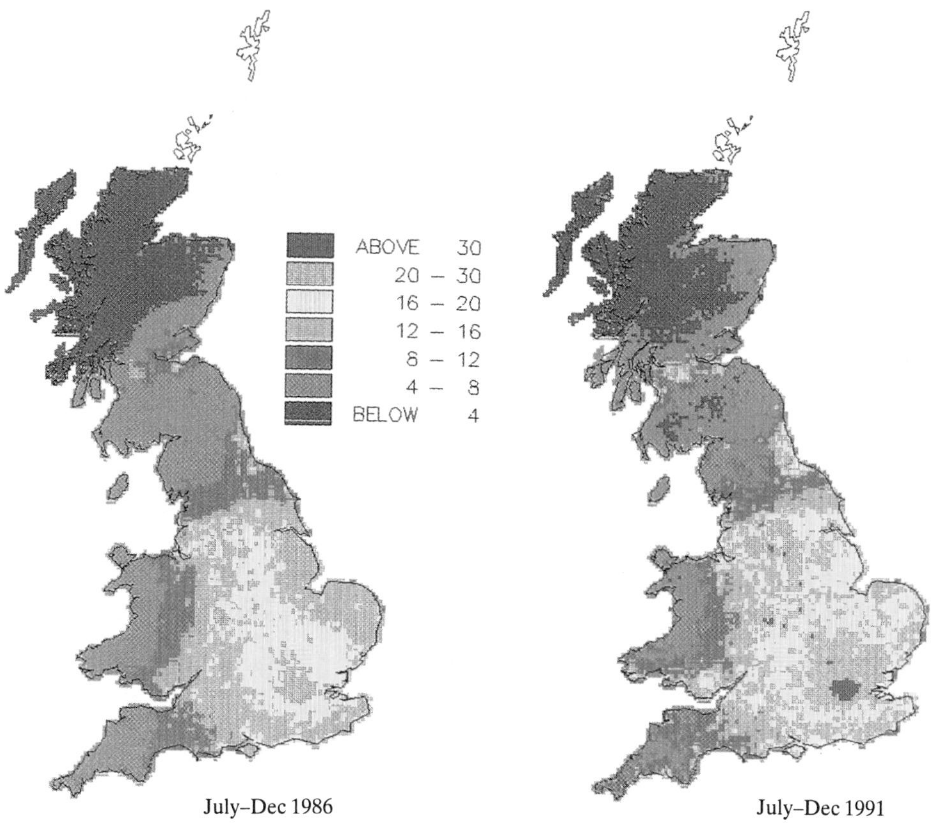

July–Dec 1986

July–Dec 1991

wheezy bronchitis at ages 16-23 (ie, in the late 1970s). A more recent nationwide survey of school children of all ages also found a slightly lower prevalence of wheezing in Scotland, but was unable to confirm the excess in southwest England or Wales.[93]

7.75   An insight into the geographical variation in asthma among young adults is provided by a very large survey of 74,000 respondents aged 20-44 years in 20 local authority districts of England in 1986.[94] The districts were selected to include those with the highest and lowest asthma mortality rates over the preceding ten years. In spite of this design, which might have been expected to maximise the range of prevalence rates obtained, the proportion of subjects reporting wheeze in the past year varied from 8.8% in Hambleton, North Yorkshire, to 13.3% in St Helens, Lancashire. The prevalence of nocturnal breathlessness (presented as an indicator of more severe asthma) varied from 2.8% in Boothferry, Humberside, to 4.8% in Bournemouth, Hampshire. Three sets of nearby areas had a very similar prevalence of wheeze: in inner London: 12.2% in Greenwich, 12.4% in Hammersmith and Fulham and 12.6% in Wandsworth; in Derbyshire: 11.3% in Bolsover, 12.0% in Erewash and 11.9% in High Peak; and in central England: 11.0% in North Warwickshire and 11.5% in Nuneaton. These results, based on large sample sizes (1,620 to 7,010 in each area) suggest a lack of fine spatial variation in asthma prevalence in England. A similar impression is gained from the results of surveys of school children. For instance, among 7-year-olds during the second half of the 1980s, the prevalence of wheeze in the past year was 12.5% in Edinburgh, 12.0% in Southampton and 11.1% in Walsall (see Table 7.1).

7.76   Geographical variations in health service utilisation may be influenced by disease prevalence or patterns of supply and demand for health care. There is very limited data available from primary care. The proportion of the population consulting their general practitioner for asthma in the 1981–82 national morbidity survey did not vary greatly with latitude.[95]

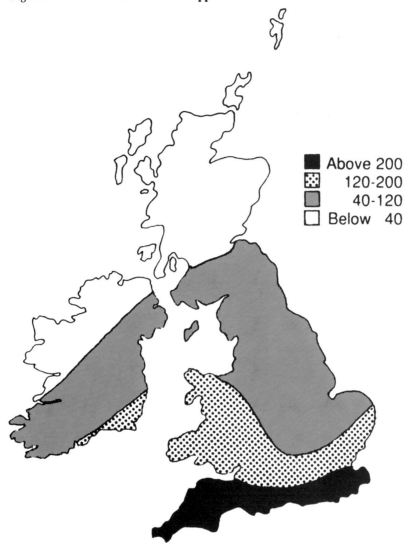

Figure 7.19c **Ozone Hours above 80ppb 1987–1990**

Above 200
120-200
40-120
Below 40

7.77 Regional hospital admission rates based on the 10% HIPE sample can be compared with greater confidence than more recent figures, as deficiencies in the coding of diagnoses in the Hospital Episode System data vary by region and from year to year. Even in the earlier data, however, there is some uncertainty, as no published statistics are available on regional differences in admission for acute bronchitis and bronchiolitis, a major alternative diagnosis for wheezing illness in young children. Figures 7.21 and 7.22 show the extent of regional differences in hospital admission rates (by health service region) for children (Figure 7.21) and young adults (Figure 7.22) during 1979–85.

There is almost a twofold variation in admission rates for children, somewhat less for adults, but no clear geographical pattern is apparent. Admission rates in Scotland (not shown in Figures 7.21 and 7.22) are substantially lower than the England and Wales average.

7.78 Asthma mortality rates for children and young adults by health service region are also shown in Figures 7.21 and 7.22. They correlate weakly with hospital admission rates for asthma in the same age-group and region over the same time period. More recent age-specific mortality rates for the period 1988–91 are included for comparison along with urban-rural variations in Table 7.4.

Figure 7.20   Prevalence of wheeze at age 5 by Regional Health Authority in 1975

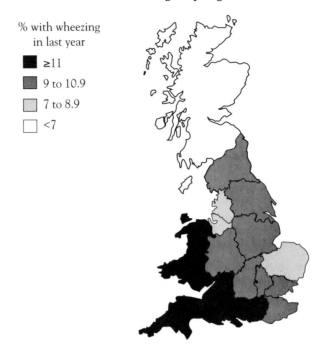

% with wheezing
in last year

■ ≥11
▨ 9 to 10.9
▢ 7 to 8.9
□ <7

Figure 7.21   Rates of hospital admission (per 10,000) and mortality (per million) for asthma by hospital region of England and Wales, for children aged 0–14 years, 1979–1985

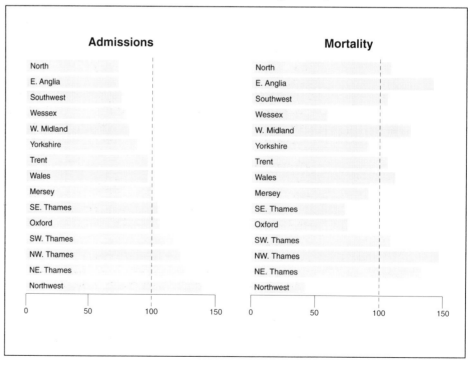

Figure 7.22 **Rates of hospital admission (per 10,000) and mortality (per million) for asthma by hospital region of England and Wales, for adults 15–44 years, 1979–1985**

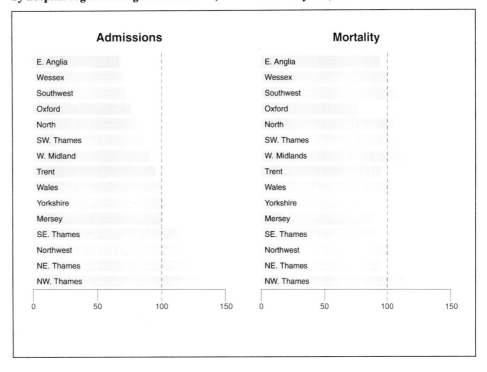

Table 7.4 **Asthma mortality rates (per million per year) by age, regional health authority and degree of urbanisation, England 1988–91**

	Age			
	0-14	**15-44**	**45-64**	**65+**
RHA				
Northern	2.2 (5)	7.1 (38)	42.4 (116)	123.9 (239)
Yorkshire	2.1 (6)	8.2 (52)	29.4 (92)	77.3 (177)
Trent	1.4 (5)	9.2 (76)	38.5 (158)	123.7 (361)
East Anglia	2.6 (4)	8.8 (31)	27.9 (49)	149.0 (206)
NW Thames	3.8 (10)	9.7 (62)	27.4 (81)	111.3 (223)
NE Thames	1.7 (5)	7.8 (53)	30.2 (96)	85.5 (198)
SE Thames	2.3 (6)	8.2 (52)	29.3 (91)	98.2 (247)
SW Thames	3.3 (7)	7.2 (37)	26.1 (68)	101.5 (204)
Wessex	1.9 (4)	9.7 (49)	27.0 (68)	119.1 (243)
Oxford	1.0 (2)	6.7 (32)	30.6 (64)	120.7 (164)
Southwestern	2.6 (6)	7.1 (39)	33.2 (95)	104.5 (247)
West Midlands	2.7 (11)	6.5 (59)	32.5 (149)	117.5 (367)
Mersey	5.4 (10)	7.2 (30)	40.0 (84)	95.9 (141)
Northwestern	3.5 (11)	8.7 (61)	44.9 (153)	102.4 (254)
Urbanisation				
Rural	1.4 (8)	7.5 (101)	28.2 (197)	115.5 (630)
Mixed	3.3 (31)	8.5 (190)	29.2 (325)	113.3 (933)
Urban	2.4 (16)	8.6 (133)	35.9 (275)	113.2 (620)
Conurbation	2.6 (37)	7.6 (247)	37.0 (567)	98.5 (1088)
All England				
Total	2.6 (92)	8.0 (671)	33.2 (1364)	108.3 (1088)

Death rates per million per year. Number of deaths in parentheses.

Other allergic diseases

7.79   A few studies using standardised (though crude) case definitions have demonstrated regional variations in the prevalence of hay fever and eczema. During 1981–82, the proportion of the population consulting their general practitioner for hay fever was higher in the south than the north of Britain. A similar north-south

111

gradient is apparent for self-reported hay fever among 12,355 members of the British 1958 cohort who were interviewed in 1981 (at age 23)[92] and for parental reports of hay fever among 13,135 5-year-olds from the British 1970 cohort surveyed in 1975.[96] The geographical pattern of eczema in early childhood in each of these two national cohorts is very similar to that of hay fever, suggesting that the north-south variation may reflect the underlying distribution of atopy in general.[92]

7.80   Studies of inter-regional migrants in the 1958 cohort (a quarter of whom had been born in a different region to their residence at 23) show that the regional differences in the prevalence of hay fever relate almost exclusively to the subjects' region of birth, and not to the region of residence at age 23 (after adjustment for region of birth).[97] This suggests that genetic or early environmental influences probably determine the north-south gradient in the prevalence of atopy in Britain.

7.81   In contrast, the regional variations in wheezing illness among young children in the 1958 and 1970 birth cohorts appeared to be more strongly related to region of residence at age 7 or 5 (respectively) than to their region of birth.[92] This suggests that the geographical pattern of wheezing in Britain, which is significantly different from that of hay fever and eczema, is determined mainly by current environmental exposures or family lifestyles.

Comment

7.82   The regional pattern of asthma within the UK is poorly defined. Population surveys and hospital admission statistics are consistent in suggesting low rates in Scotland, although possible differences in disease labelling and diagnosis between England and Scotland have not been thoroughly evaluated. There is also a suggestion that hay fever and eczema are less common in the north, so the Scots may be a group at low risk of all forms of allergy.

7.83   Within England and Wales, the pattern of asthma prevalence bears little relationship to regional variations in the main pollutants of current concern. Rates of hospital admission and mortality show no clear geographical trends and probably depend more on patterns of medical care (which may vary between closely adjacent regions) than environmental factors.

Urban-rural variations   *Air pollution*

7.84   Urban-rural variations in past and current air pollution levels have been discussed earlier under time trends and regional variations (see Figure 7.19 and Table 7A.7 in Annex 7A). Up to the 1960s, there was an urban excess of particulate and SO_2 pollution. Nowadays, particulate and NO_x levels are higher in urban areas, ozone levels are higher in rural areas, and there is little urban-rural variation in levels of SO_2.

Asthma

7.85   Remarkably little has been published on urban-rural variations in asthma prevalence within the UK. Routine statistics relating to hospital admissions and general practitioner consultations are not presented in this way. The material to be presented in this section is derived largely from previously unpublished analyses of prevalence data from the 1958 and 1970 national birth cohorts, classified by local authority area of birth, and mortality data for England and Wales for the period 1988-91, classified by health district of residence.

7.86   Table 7.5 shows the prevalence of wheezing illness among children in the two national cohorts, and the parents of the 1970 cohort, by the child's area of birth classified by degree of urbanisation.

Table 7.5 **Prevalence (%) of asthma and wheezing illness by degree of urbanisation in the British 1958 and 1970 birth cohorts and parents of the 1970 cohort**

1958 Cohort	Conurbation	County Borough	Urban County	Rural County
Asthma or wheezy bronchitis ever by age 7	19.4 793/4091	17.7 506/2858	17.8 620/3481	18.7 673/3596
Asthma or wheezy bronchitis in the past year at 7	8.8 358/4091	7.1 204/2858	8.6 299/3481	8.2 294/3596
5 or more attacks of asthma or WB in past year at 7	1.0 39/4091	0.8 23/2858	0.9 30/3003	0.9 33/3596
Asthma or wheezy bronchitis at ages 16-23	9.4 312/3312	9.3 227/2450	10.2 306/3003	9.1 281/3106
1970 Cohort				
Wheezing ever by age 5	21.4 642/2999	20.3 419/2064	19.7 755/3830	21.8 785/3609
Wheezing in the first year of life	8.5 255/2999	8.0 165/2064	8.1 312/3830	8.2 299/3609
Wheezing in the past year at 5	9.1 274/2999	9.6 198/2064	9.7 371/3830	9.4 339/3609
5 or more attacks of wheezing in past year at 5	1.0 31/2999	1.2 24/2064	0.9 34/3830	1.0 37/3609
Mother ever had asthma	4.2 111/2620	4.2 75/1796	4.4 151/3408	4.8 156/3173
Father ever had asthma	3.8 96/2507	5.9 102/1737	4.6 151/3318	5.4 168/3111

No substantial urban-rural variation in prevalence or frequency of wheezing was apparent in either cohort, at any age up to 23, nor among the parents of the 1970 birth cohort (aged approximately 20-40 in 1975). Those who were born in urban areas in 1958 were exposed during the first five years of life to annual mean levels of smoke and SO_2 pollution which were at least three times greater than those encountered in urban areas today (see Figure 7.11), and to winter peaks of pollution which regularly exceeded 500 $\mu g/m^3$ Black Smoke. Such exposures would have been extremely uncommon for children born in urban areas in 1970.

7.87   A recent nationwide prevalence survey of school children of all ages in selected British parliamentary constituencies found a similar annual period prevalence of wheezing in urban (15.3%), mixed urban/rural (15.0%) and rural constituencies (14.1%), but frequent wheeze was about half as common in the rural constituencies as in urban or mixed constituencies.[93] This finding was based on relatively small numbers of children in the rural communities and its statistical significance is difficult to evaluate because of the technique of cluster sampling (by constituency). A recent survey of children in the Scottish island of Skye generated prevalence figures comparable with those obtained elsewhere in the UK (see Table 7.1), despite the low levels of motor traffic, no industrial pollution sources and presumably low ozone levels (on account of latitude).[98]

7.88   Urban-rural differences in mortality rates for asthma are shown in Table 7.4. There is little evidence of an urban excess of mortality among children and young adults, suggesting that disease severity does not vary greatly between town and country. These figures for middle-aged and older adults need to be interpreted with caution, as there is great potential for diagnostic confusion with chronic bronchitis, emphysema and obstructive airways disease, causes of death which are known to be much more common in urban areas.

7.89   A similar analysis of hospital admission rates in urban and rural areas is not possible at present, due to deficiencies in the coding of diagnosis and area of residence in the Hospital Episode System data for England since 1987. Earlier (10% HIPE) data were not disaggregated to district level.

Other allergic diseases

7.90   The prevalence of self-reported hay fever among young adults from the 1958 cohort who were living in British conurbations in the late 1970s and early 1980s differed little from the prevalence reported in their respective regional hinterlands.[97] The proportion reporting hay fever in the past year at age 23 was 16% in the conurbations (Greater London, metropolitan counties and Greater Glasgow) and 17% in other areas.

7.91   During the 1980s, consultations for hay fever were about 20% more frequent in urban general practices reporting episodes of selected acute illnesses to the Royal College of General Practitioners, although a similar urban excess was also apparent for other acute respiratory diseases.[62] The seasonality of hay fever was very similar in urban and rural areas, and no divergence between urban and rural consulting rates was evident over the period 1981–92.

Comment

7.92   Despite substantial urban-rural variations in past and current levels of air pollution, there is little evidence of variation in asthma prevalence or mortality by degree of urbanisation, either in the generations exposed to "old-fashioned" smoke and SO_2 pollution, or among children and adults exposed to traffic in modern towns and cities. The limited information available does not suggest a marked urban excess of allergic rhinitis in the UK.

7.93   Other studies in developed countries are consistent with these observations. A large and carefully designed population survey of chronic respiratory disease was carried out among men and women aged 30-64 years in Uppsala and the surrounding areas in 1966–67.[99] The prevalence of asthma, defined in a self-completed questionnaire as "wheezing and difficulty in breathing", did not differ between 34,092 subjects from urban Uppsala (2.3%) and 7,587 from the surrounding rural areas (2.3%). In a recent study of childhood asthma in Victoria, Australia, the prevalence of wheeze in the past year was found to be similar in five rural areas as in the city of Melbourne.[100] However, a significant variation in prevalence among the rural areas was found, which could not be explained by known environmental exposures.

7.94   In contrast, two carefully conducted studies of children in Zimbabwe[101] and South Africa[102] have found marked differences in the prevalence of exercise-induced bronchospasm between urban and rural communities of similar genetic origin. In these rural samples there was virtually no exercise-induced bronchospasm, whereas the urban children had rates comparable to those obtained in developed countries. There is also evidence of asthma emerging as a problem among Pacific island communities who were relocated to New Zealand.[103] This suggests that the conditions conducive to the development of childhood asthma may be those associated with the "Western" culture in general, rather than with urban living itself.

Conclusions

7.95   Trends in wheezing illness in the UK have generally been upward over the past thirty years. However, the increase in disease prevalence, among school children (the only group with adequate data) appears to have been modest, and there is a suggestion that it may now have levelled off. Indicators of health service activity have shown much larger proportional increases, particularly among the very young, but there is good reason to believe that these trends exaggerate changes in the prevalence and severity of asthma, because of documented shifts in the patterns of diagnosis, clinical management and use of health services. The most recent hospital admission data suggest that the upward trend has now ceased and may even have reversed in some areas. There is thus uncertainty about whether asthma is increasing in prevalence or severity nowadays, and there is suggestive evidence that past trends in prevalence may have been influenced by an increase in the prevalence of atopy. Increasing exposure to air pollution is only one of several explanations which have been advanced to explain such a rise in allergic sensitisation.

7.96   Asthma is a disease characterised by considerable short-term variability due to factors other than air pollution: epidemics of virus infection; aeroallergen exposure; and perhaps abrupt climate changes. Marked variations in the occurrence of asthma can thus occur without a corresponding variation in pollution level. It is, therefore, important to control as rigorously as possible for regular seasonal cycles and irregular variations of known cause (such as epidemics of virus infections including influenza), before evaluating the effect of air pollution on asthma incidence. If such adjustment is not done, then the effect of pollution peaks might be obscured by contrary fluctuations in other factors, or exaggerated by a coincidental rise in pollution levels at the same time as other influences begin to operate. Such adjustments require complex statistical techniques which are most readily applied to analysis of complete time-series of daily admissions or attendances at health care facilities (see the section on time-series in Chapter 8).

7.97   In contrast, the spatial distribution of asthma appears remarkably uniform within developed countries. No urban-rural differences in asthma prevalence are apparent in the UK, despite major urban-rural gradients in the levels of past and current air pollution. Within England and Wales, the pattern of asthma prevalence bears little relationship to regional variations in the main pollutants of current concern. Hospital admission rates and mortality are not clearly related to geographical variables and probably depend more on patterns of medical care than environmental factors.

References

1.   Luyt DK, Burton PR, Simpson H. Epidemiological study of wheeze, doctor diagnosed asthma and cough in preschool children in Leicestershire. BMJ 1993; 306:1386-1390.

2.   Anderson HR. Is asthma really increasing? Paediatr Respir Med 1993; 1:6-10.

3.   Lenney W, Wells NEJ, O'Neill BA. The burden of paediatric asthma. Eur Respir Rev 1994:4:49-62.

4.   Dawson B, Horobin G, Illsley R, Mitchell RG. A survey of childhood asthma in Aberdeen. Lancet 1969; i:827-830.

5.   Graham PJ, Rutter ML, Yule W, Pless IB. Childhood asthma. A psychosomatic disorder? Some epidemiological considerations. Br J Prev Soc Med 1967; 21:78-85.

6.   Anderson HR, Bland JM, Patel S, Peckham C. The natural history of asthma in childhood. J Epidemiol Community Health 1986; 40:121-129.

7.   Hamman RF, Halil T, Holland WW. Asthma in school children: demographic associations and peak expiratory flow rates compared to children with bronchitis. Br J Prev Soc Med 1975; 24:228-238.

8.   Leeder SR, Corkhill R, Irwig LM, Holland WW, Corkhill R. Influence of family factors on the incidence of lower respiratory illness during the first five years of life. Br J Prev Soc Med 1976; 30:302-312.

9.   Smith JM, Harding LK, Cumming G. The changing prevalence of asthma in schoolchildren. Clin Allergy 1971; 1:57-61.

10.   Peckham C, Butler N. A national study of asthma in childhood. J Epidemiol Community Health 1978; 32:79-85.

11.   Burr ML, Butland BK, King S, Vaughan-Williams E. Changes in asthma prevalence: two surveys 15 years apart. Arch Dis Child 1989; 64:1425-1456.

12.   Smith JM. The prevalence of asthma and wheezing in children. Br J Dis Chest 1976; 70:73-77.

13.   Golding J, Butler NR. Wheezing and asthma. In: Butler NR, Golding J, editors. From birth to five: a study of the health and behaviour of Britain's 5-year-olds. Oxford: Pergamon Press, 1986; 158-170.

14. Strachan DP, Anderson HR. Trends in hospital admission rates for asthma in children. BMJ 1992; 304:819-820.

15. Lee DA, Winslow NR, Speight ANP, Hey EN. Prevalence and spectrum of asthma in childhood. BMJ 1983; 286:1256-1258.

16. Colver AF. Community campaign against asthma. Arch Dis Child 1990; 59:449-452.

17. Johnston ID, Bland JM, Anderson HR. Ethnic variation in respiratory morbidity and lung function in childhood. Thorax 1987; 42:542-548.

18. Hill R, Williams J, Tattersfield A, Britton J. Change in the use of asthma as a diagnostic label for wheezing illness in children. BMJ 1989; 299:898.

19. Strachan DP. Damp housing and childhood asthma: validation of reporting of symptoms. BMJ 1988; 297:1223-1226.

20. Clifford RD, Radford M, Howell JB, Holgate ST. Prevalence of respiratory symptoms among 7 and 11 year old schoolchildren and association with asthma. Arch Dis Child 1989; 64:1118-1125.

21. Ninan TK, Russell G. Respiratory symptoms and atopy in Aberdeen schoolchildren: evidence from two surveys 25 years apart. BMJ 1992; 304:873-875.

22. Symington P, Coggon D, Holgate S. Respiratory symptoms in children at schools near a foundry. Br J Ind Med 1991; 48:588-591.

23. Strachan DP, Anderson HR, Limb ES, O'Neill A, Wells N. A national survey of asthma prevalence, severity and treatment in Great Britain. Arch Dis Child 1994; 70:174-178.

24. Lewis S, Butland B, Strachan D, Richards D, Butler N, Bynner J, Britton J. A comparison of the prevalence and severity of childhood wheezing illness in 1974 and 1986 using two nationally representative British born birth cohorts [Abstract]. Am J Respir Crit Care Med 1994; 149(4 Part 2):A574.

25. Burney PG, Chinn S, Rona RJ. Has the prevalence of asthma increased in children? Evidence from the national study of health and growth. BMJ 1990; 300:1306-1310.

26. Whincup PH, Cook DG, Strachan DP, Papacosta O. Time trends in respiratory symptoms in childhood over a 24 year period. Arch Dis Child 1993; 68:729-734.

27. Cook DG, Shaper AG. The respiratory benefits of stopping smoking. J Smoking Rel Dis 1990; 1:45-58.

28. Burr ML. Epidemiology of asthma. In: Burr ML, editor. Epidemiology of clinical allergy. Basel: Karger, 1993; 80-102.

29. Robertson CF, Heycock E, Bishop J, Nolan T, Olinsky A, Phelan PD. Prevalence of asthma in Melbourne schoolchildren: changes over 26 years. BMJ 1991; 302:1116-1118.

30. Varioner HS, de Haller J, Schopfer C. Prévalence de l'allergie chez les enfants et les adolescents. Helv Paediatr Acta 1984; 39:129-136.

31. Mitchell EA. Increasing prevalence of asthma in children. N Z Med J 1983; 96:463-464.

32. Shaw RA, Crane J, O'Donell TV, Porteous LE, Coleman ED. Increasing asthma prevalence in a rural New Zealand adolescent population: 1975–1989. Arch Dis Child 1990; 65:1319-1323.

33. Weitzman M, Gortmaker SL, Sobol AM, Perrin JM. Recent trends in the prevalence and severity of childhood asthma. JAMA 1992; 268:2673-2677.

34. Hsieh KH, Shen JJ. Prevalence of childhood asthma in Taipei, Taiwan and other Asian Pacific countries. J Asthma 1988; 25:73-82.

35. Aberg N. Asthma and allergic rhinitis in Swedish conscripts. Clin Exp Allergy 1989; 19:59-63.

36. Haahtela T, Lindholm H, Bjorkstein F, Koskenvuo K, Laitenen LA. Prevalence of asthma in Finnish young men. BMJ 1990; 301; 266-268.

37. Laor A, Cohen L, Danon YL. Effects of time, sex, ethnic origin and area of residence on prevalence of asthma in Israeli adolescents. BMJ 1993; 307:841-844.

38. Peat JK, Haby M, Spijker J, Berry G, Woolcock AJ. Prevalence of asthma in adults in Busselton, Western Australia. BMJ 1992; 305:1326-1329.

39. Baumann A. Has the prevalence of asthma symptoms increased in Australian children? J Paediatr Child Health 1993; 29:424-428.

40. Anderson HR, Butland BK, Strachan DP. Trends in the morbidity and disability of wheezy children in Croydon 1978–1991. BMJ 1994 [In Press]

41. Anderson HR, Butland BK, Paine KM, Strachan DP. Trends in the medical care of asthma in childhood: Croydon 1978 and 1991 [Abstract]. Thorax 1993; 48:P451.

42. Anderson HR. Increase in hospitalisation for childhood asthma. Arch Dis Child 1978; 53:295-300.

43. Alderson MR. Trends in morbidity and mortality from asthma. Popul Trends 1987; 49:18-23.

44. Anderson HR. Increase in hospital admissions for childhood asthma: trends in referral, severity and readmissions from 1970 to 1985 in a health region of the United Kingdom. Thorax 1989; 44:614-619.

45. Hill AM. Trends in paedriatric medical admissions. BMJ 1989; 298:1479-1483.

46. Storr J, Barrell E, Lenney W. Rising asthma admissions and self- referral. Arch Dis Child 1988; 63:774-779.

47. Anderson HR, Freeling P, Patel SP. Decision-making in acute asthma. J R Coll Gen Pract 1983; 33:105-108.

48. Mackay TW, Wathen CG, Sudlow MF, Elton RA, Caulton E. Factors affecting asthma mortality in Scotland. Scot Med J 1992; 37:5-7.

49. Hyndman S, Williams DRR. Falling asthma admission rates: artifact or reality? BMJ 1994 [In press]

50. Fleming DM, Crombie DL. Prevalence of asthma and hay fever in England and Wales. BMJ 1987; 294:279-283.

51. Ayres JG, Noah ND, Fleming DM. Incidences of episodes of acute asthma and acute bronchitis in general practice 1976–1987. Br J Gen Pract 1993; 43:361-364.

52. Weiss KB, Gergen PJ, Wagener DK. Breathing better or wheezing worse? The changing epidemiology of asthma morbidity and mortality [Review]. Annu Rev Public Health 1993; 14:491-513.

53. Mao Y, Semenciw R, Morrison H, MacWilliam L, Davies J, Wigle D. Increased rates of illness and death from asthma in Canada. Can Med Assoc J 1987; 137:620-623.

54. Mitchell EA. International trends in hospital admission rates for asthma. Arch Dis Child 1985; 60:376-378.

55. Inman WHW, Adelstein AM. Rise and fall of asthma mortality in England and Wales in relation to use of pressurised aerosols. Lancet 1969; ii:279-283.

56. Burney PGJ. Asthma deaths in England and Wales 1931–1985: evidence for a true increase in asthma mortality. J Epidemiol Community Health 1988; 42:316-320.

57. Burney PGJ. Asthma: evidence for a rising prevalence. Proc R Coll Phys Edin 1993; 23:595-600.

58. Emmanuel MB. Hayfever, a post industrial revolution epidemic: a history of its growth during the 19th century. Clin Allergy 1988; 18:295-304.

59. Wüthrich B. Epidemiology of the allergic diseases: are they really on the increase? Int Arch Allergy Appl Immunol 1989; 90:3-10.

60. Hagy GW, Settpinae GA. Bronchial asthma, allergic rhinitis and allergy skin tests among college students. J Allergy 1969; 44:323-332.

61. Aberg N, Hesselmar B, Aberg B, Eriksson B. Increase of asthma, allergic rhinitis and eczema in Swedish school children between 1979 and 1991. Clin Exp Allergy 1995 [In press].

62. Ross AM, Fleming DM. Incidence of allergic rhinitis in general practice 1981–92. BMJ 1994; 308:897-900.

63. Taylor B, Wadsworth M, Wadsworth J, Peckham C. Changes in the reported prevalence of childhood eczema since the 1939–45 war. Lancet 1984; ii:1255-1257.

64. Schultz-Larsen F, Holm NV, Henningsen K. Atopic dermatitis. A genetic-epidemiological study in a population-based twin sample. J Am Acad Dermatol 1986; 15:487-494.

65. Barbee RA, Kaltenborn W, Lebowitz MD, Burrows B. Longitudinal changes in allergen skin test reactivity in a community population sample. J Allergy Clin Immunol 1987; 79:16-24.

66. Sibbald B, Rink E, D'Souza M. Is atopy increasing? Br J Gen Pract 1990; 40:338-340.

67. Nakagomi T, Itaya H, Tominaga T, Yamaki M, Hisamatsu S, Nakagomi N. Is atopy increasing? Lancet 1994; 343:121-122.

68. Quality of Urban Air Review Group. Urban Air Quality in the United Kingdom. First Report. Bradford: Department of the Environment, 1993.

69. Quality of the Urban Air Review Group. Diesel vehicle emissions and urban air quality. Second Report. Birmingham: Department of the Environment, 1993.

70. Department of Health. Advisory Group on the Medical Aspects of Air Pollution Episodes. First Report: Ozone. London: HMSO; 1991.

71. Department of Health. Advisory Group on the Medical Aspects of Air Pollution Episodes. Third Report: Oxides of Nitrogen. London: HMSO, 1993.

72. Bower JS, Lampert JE, Stevenson KJ, Atkins DHF, Law DV. A diffusion tube survey of NO_2 levels in urban areas of the UK. Atmos Environ 1991; 25B:255-265.

73. Ayres JG. Trends in asthma and hay fever in general practice in the United Kingdom 1976–83. Thorax 1986; 41:110-116.

74. Ayres JG. Seasonal pattern of acute bronchitis in general practice in the United Kingdom 1976–83. Thorax 1986; 41:106-110.

75. Khot A, Burn R, Evans N, Lenney C, Lenney W. Seasonal variation and time trends in childhood asthma in England and Wales 1975–1981. BMJ 1984; 289:235-237.

76. Khot A, Burn R. Seasonal variation and time trends of deaths from asthma in England and Wales 1960–82. BMJ 1984; 289:233-234.

77. Storr J, Lenney W. School holidays and admissions with asthma. Arch Dis Child 1989; 64:103-107.

78. Pattemore PK, Johnston SL, Bardin PG. Viruses as precipitants of asthma symptoms. I: Epidemiology. Clin Exp Allergy 1992; 22:325-336.

79. Horn MEC, Brain EA, Gregg I, Inglis JM, Yealland SJ, Taylor P. Respiratory viral infection and wheezy bronchitis in childhood. Thorax 1979; 34:23-28.

80. Hudgel DW, Langston L, Selner JC, McIntosh K. Viral and bacterial infections in adults with chronic asthma. Am Rev Respir Dis 1979; 120:393-397.

81. Carlsen KH, Orstavik I, Leegaard J, Hoeg H. Respiratory virus infections and aeroallergens and acute bronchial asthma. Arch Dis Child 1984; 59:310-315.

82. Fleming DM, Crombie DL, Norbury CA, Cross KW. Observations on the influenza epidemic of November/December 1989. Br J Gen Pract 1990; 40:495-497.

83. Greenburg L, Field F, Reed JI, Erhardt CL. Asthma and temperature change. Arch Environ Health 1964; 8:643-647.

84. Greenburg L, Field F, Reed JI, Erhardt CL. Asthma and temperature change II. 1964 and 1965 epidemiological studies of emergency visits for asthma in three large New York City hospitals. Arch Environ Health 1966; 12:561-563.

85. Buckland FE, Tyrell DAJ. Loss of infectivity on drying various viruses. Nature 1962; 195:1063-1064.

86. Green HG. The health implications of the levels of indoor air humidity. In: Berglund B, Lindvall T, Sundell J, editors. Indoor Air. Proceedings of the 3rd international conference on indoor air quality and climate. Volume 3. Sensory and hyperreactivity reactions to sick buildings. Stockholm: Swedish Council for Building Research, 1984; 71-77.

87. Kingdom KH. Relative humidity and airborne infections. Am Rev Respir Dis 1960; 81:504-512.

88. Lester W. The influence of relative humidity on the infectivity of airborne influenza A virus. J Exp Med 1948; 88:361-368.

89. Anon. Asthma and the weather [Editorial]. Lancet 1985; i:1079-1080.

90. Carey MJ, Cordon I. Asthma and climatic conditions: experience from Bermuda, an isolated island community. BMJ 1986; 293:843-844.

91. Eggleston S, Hackman MP, Heyes CA, Irwin JG, Timmis RJ, Williams ML. Trends in urban air pollution in the United Kingdom during recent decades. Atmos Environ 1992; 26B:227-240.

92. Strachan DP, Golding J, Anderson HR. Regional variations in wheezing illness in British children: the effect of migration during early childhood. J Epidemiol Community Health 1990; 44:231-236.

93. Strachan DP, Anderson HR, Limb ES, O'Neill A, Wells N. A national survey of asthma prevalence, severity and treatment in Great Britain. Arch Dis Child 1994; 70:174-178.

94. Burney PGJ, Papacosta AO, Withey CH, Holland WW, Colley JRT. Hospital admission rates and the prevalence of asthma symptoms in 20 local authority districts. Thorax 1991; 46:574-579.

95. Fleming DM, Crombie DL. Geographical variations in persons consulting rates in general practice in England and Wales. Health Trends 1989; 21:51-55.

96. Golding J, Peters T. Eczema and hay fever. In: Butler NR, Golding J, editors. From birth to five. A study of the health and behaviour of Britain's five-year-olds. Oxford: Pergamon Press, 1986; 171-186.

97. Sibbald B, Strachan D. Epidemiology of rhinitis. In: Holgate ST, Busse WW, editors. Mechanisms in asthma and rhinitis: implications for diagnosis and treatment. Oxford: Blackwell Scientific Publications, 1994 [In press].

98. Austin JB, Russell G, Adam MG, Macintosh D, Kelsey S, Peck DF. Prevalence of asthma and wheeze in the highlands of Scotland. Arch Dis Child 1994; 71:211-216.

99. Irnell L, Kivikoog J. Bronchial asthma and chronic bronchitis in a Swedish urban and rural population. With special reference to prevalence, respiratory function and socio-medical condition. Scand J Respir Dis 1968; 66(Suppl):1-86.

100. Robertson CF, Bishop J, Dalton M, Caust J, Nolan TM, Olinsky A, Phelan PD. Prevalence of asthma in regional Victorian schoolchildren. Med J Australia 1992; 156:831-833.

101. Keeley DJ, Neill P, Gallivan S. Comparison of the prevalence of reversible airways obstruction in rural and urban Zimbabwean children. Thorax 1991; 46:549-553.

102. van Niekerk CH, Weinberg EG, Shore SC, Heese H de V, van Schalkwyk DJ. Prevalence of asthma: a comparative study of urban and rural Xhosa children. Clin Allergy 1979; 9:319-324.

103. Waite DA, Eyles EF, Tonkin SL, O'Donnell TV. Asthma prevalence in Tokelauan children in two environments. Clin Allergy 1980; 10:71-75.

Table 7A.1 **Mean weekly attack rates for acute asthma (per 100,000 per week) recorded by RCGP Weekly Return Service, both sexes, by age, England, Scotland and Wales, 1976–92**

Year	0-4	5-14	15-44	45-64	65+
1976	13.5	17.4	8.9	9.0	6.4
1977	13.7	15.2	8.9	10.2	8.3
1978	26.6	23.4	11.7	12.2	10.5
1979	33.0	25.6	13.4	13.1	11.1
1980	31.6	28.8	15.0	14.5	12.9
1981	31.3	32.7	15.1	13.8	11.7
1982	32.2	31.5	15.6	13.7	13.3
1983	35.4	35.7	13.4	11.4	10.7
1984	52.4	36.5	13.7	12.5	10.0
1985	57.1	43.3	16.2	14.4	10.2
1986	52.4	41.4	14.4	12.8	11.8
1987	74.4	58.9	20.6	17.1	13.4
1988	102.4	68.7	24.0	20.8	18.9
1989	92.9	66.3	24.8	23.6	22.9
1990	108.4	71.0	26.9	26.2	25.0
1991	145.9	89.4	33.5	29.5	28.8
1992	147.8	84.2	36.6	31.4	32.3

Sources: Ayres JG, Noah ND, Fleming DM. Br J Gen Pract 1993; 43:361-364 (1976–87) and *ad hoc* tabulations by RCGP Research Unit, Birmingham (1988–1992)

Table 7A.2 **Persons consulting for asthma in general practice (per 1,000 per year) in national morbidity surveys, 1970–1, 1981–2 and 1991–2**

Year	0-4	5-14	15-24	25-44	45-64	65-74	75+
Both sexes							
1970/1	12.9	16.0	9.5	8.3	9.0	9.8	6.0
1981/2	26.0	29.3	16.6	12.1	16.1	19.9	14.0
1991/2	86.1	75.5	42.8	29.6	30.0	39.4	31.2
Males							
1970/1	16.9	21.9	10.8	8.0	8.1	10.7	6.6
1981/2	33.3	37.5	18.6	12.1	13.9	21.5	16.2
1991/2	99.4	86.0	39.6	25.8	26.0	37.2	34.5
Females							
1970/1	8.7	9.7	8.3	8.6	9.7	9.1	5.7
1981/2	18.3	20.5	14.7	12.1	18.1	18.7	12.9
1991/2	72.2	64.5	45.9	33.4	34.2	41.2	29.5

Sources: RCGP, OPCS and DHSS/DH. Morbidity statistics from general practice: 2nd, 3rd and 4th national studies

Table 7A.3 **Hospital admission rates for asthma (per 10,000 per year) by age and sex, England and Wales, 1958–85**

Year	Males					Females				
	0–4	5–14	15–44	45–64	65+	0–4	5–14	15–44	45–64	65+
1958	6.70	5.00	2.90	4.00	3.40	3.90	2.70	4.20	5.40	3.50
1959	5.60	4.40	2.00	4.00	3.30	3.00	2.90	3.70	5.30	3.30
1960	6.00	4.00	2.10	4.10	3.60	4.70	2.90	4.00	6.00	4.10
1961	5.90	5.40	2.50	4.30	3.60	5.80	2.80	4.20	5.80	4.20
1962	5.00	4.39	2.22	4.11	4.23	3.01	3.10	4.45	6.01	3.90
1963	6.07	6.04	3.14	5.62	4.57	5.23	3.40	4.78	7.49	4.78
1964	6.22	7.27	3.62	6.21	4.13	5.23	4.77	5.64	8.34	4.32
1965	8.90	7.58	4.22	7.00	5.70	6.06	5.33	5.86	9.02	6.06
1966	8.50	9.05	3.99	7.12	6.60	4.83	6.44	6.13	8.70	5.68
1967	11.65	10.52	4.98	8.09	6.14	8.05	7.14	7.05	9.11	6.19
1968	11.66	12.89	4.82	8.17	8.08	8.15	7.26	7.21	9.34	6.74
1969	13.59	11.26	4.96	7.26	7.75	8.90	6.39	7.26	9.21	6.52
1970	13.66	11.44	3.61	6.99	6.44	8.27	5.52	6.06	7.79	5.88
1971	16.14	11.12	4.15	6.30	5.74	9.32	7.84	6.15	7.97	6.03
1972	14.39	10.59	3.51	6.81	6.42	7.84	6.92	6.04	8.26	6.52
1973	19.11	12.68	3.58	6.03	6.07	11.75	7.20	5.69	7.86	6.71
1974	15.19	10.90	3.40	5.70	6.19	11.78	5.96	5.43	7.34	7.04
1975	19.29	13.23	3.43	5.76	5.52	12.04	7.41	5.42	7.56	5.91
1976	22.56	12.73	3.78	6.40	5.96	15.52	8.53	5.80	7.82	6.61
1977	25.71	15.52	4.12	6.29	6.14	15.96	8.92	6.16	7.59	6.10
1978	30.93	16.99	4.16	6.07	6.96	20.87	9.64	6.49	7.81	6.64
1979	32.06	17.62	4.13	6.25	6.44	20.80	10.53	6.49	7.69	6.79
1980	41.49	19.87	4.63	6.67	6.59	26.10	13.02	6.40	8.13	7.65
1981	57.31	25.00	5.42	7.33	7.46	28.98	15.03	7.95	8.80	8.42
1982	70.02	27.83	5.24	8.31	7.59	35.49	16.70	8.06	9.28	9.38
1983	75.60	31.13	5.73	8.35	8.68	42.22	18.67	8.47	10.28	8.89
1984	94.01	33.41	5.72	9.22	9.78	48.45	20.05	9.26	10.65	10.83
1985	100.58	34.99	6.88	10.39	9.46	49.25	20.71	10.32	12.73	10.93

Source: Hospital Inpatient Enquiry plus *ad hoc* requests of Welsh data for 1982–85.

Table 7A.4 **Hospital admission rates for asthma (per 10,000 per year) by age, sex and country, 1982–93**

	Males						Females					
England	**0–4**	**5–14**	**15–44**	**45–64**	**65–74**	**75+**	**0–4**	**5–14**	**15–44**	**45–64**	**65–74**	**75+**
1982	70.6	27.7	5.2	8.4	8.2	6.2	35.9	16.8	8.0	9.2	10.2	8.0
1983	76.1	30.9	5.7	8.3	10.0	5.9	42.8	18.8	8.4	10.1	9.5	7.4
1984	94.3	33.6	5.7	9.1	10.4	8.1	48.5	20.0	9.3	10.4	11.7	9.3
1985	101.5	34.7	6.9	10.3	9.2	9.1	49.7	20.7	10.3	12.5	11.5	9.7
1986	no data	no data	no data	no data	no data	no data	no data	no data	no data	no data	no data	no data
1987	no data	no data	no data	no data	no data	no data	no data	no data	no data	no data	no data	no data
1988	132.0	40.7	7.3	9.7	9.6	7.1	69.3	24.4	13.0	12.7	12.6	9.7
1989	120.3	36.5	7.3	10.0	10.1	8.4	64.5	23.0	13.8	13.4	13.3	9.9
1990	133.2	36.3	7.2	9.0	9.3	8.3	70.5	23.3	13.9	12.7	12.6	10.0
1991	124.0	35.6	7.8	8.3	9.7	8.5	63.3	23.5	14.2	12.7	12.6	11.0
1992	120.5	30.9	7.5	8.7	9.8	8.2	60.6	20.2	13.8	12.9	13.6	11.0
1993	122.2	31.8	7.7	9.3	11.2	10.2	63.6	21.5	15.9	14.2	15.9	12.8
Scotland	**0–4**	**5–14**	**15–44**	**45–64**	**65–74**	**75+**	**0–4**	**5–14**	**15–44**	**45–64**	**65–74**	**75+**
1982	40.7	17.5	5.7	8.2	7.9	6.1	27.3	9.1	7.9	10.3	9.6	8.0
1983	50.1	24.2	5.8	8.8	10.0	8.2	30.7	11.7	8.6	10.2	10.7	9.6
1984	69.9	22.7	5.6	9.1	9.8	8.5	39.5	11.7	9.0	10.4	10.3	10.3
1985	66.5	24.0	6.1	8.7	9.8	8.5	41.0	14.6	9.6	11.5	11.0	9.2
1986	72.6	23.8	5.9	9.6	9.9	9.8	42.6	15.1	10.0	11.4	10.9	10.3
1987	97.0	31.2	6.7	9.4	10.4	12.3	56.1	19.7	11.2	11.7	11.4	10.5
1988	104.7	29.5	7.3	10.3	12.0	9.9	54.3	18.2	11.7	14.6	11.5	11.6
1989	92.7	30.1	7.0	10.7	10.8	10.6	49.3	18.5	14.2	15.6	12.2	10.4
1990	121.2	36.3	7.6	8.3	10.6	9.7	64.2	21.4	14.7	13.7	13.2	11.0
1991	129.2	41.0	9.3	9.7	9.7	10.5	55.8	24.4	15.6	13.9	11.3	12.0
1992	123.6	32.6	8.9	10.5	10.9	10.5	62.3	21.0	15.5	14.1	12.7	11.7
1993	124.0	34.5	9.6	9.4	9.9	11.8	64.5	23.5	18.6	16.7	15.4	11.7
Wales	**0–4**	**5–14**	**15–44**	**45–64**	**65–74**	**75+**	**0–4**	**5–14**	**15–44**	**45–64**	**65–74**	**75+**
1982	58.1	28.6	5.1	7.1	9.5	5.5	27.8	14.7	8.7	11.1	12.0	9.8
1983	64.7	33.2	5.3	8.5	9.6	10.1	30.6	16.1	8.9	12.9	16.4	12.7
1984	86.5	28.9	4.9	10.8	13.9	11.1	46.0	19.5	8.8	14.2	16.6	12.1
1985	83.6	36.7	6.4	13.0	12.9	14.9	39.4	19.7	9.8	17.0	15.3	15.3
1986	91.4	37.7	7.6	12.3	13.6	13.3	45.6	23.2	11.6	16.4	21.0	16.1
1987	105.8	40.3	7.4	12.9	14.6	11.1	50.8	25.6	11.2	16.4	16.9	16.4
1988	121.4	40.8	6.8	13.3	13.4	14.6	63.9	22.2	13.3	18.1	19.6	17.7
1989	101.0	31.6	6.3	10.9	14.8	13.5	52.4	18.2	12.7	18.7	19.4	16.4
1990	88.3	27.8	6.0	9.9	12.2	9.5	55.8	19.1	11.3	15.4	20.0	12.4

Sources: English data estimated from Hospital Inpatient Enquiry (10% sample 1982–85) and Hospital Episode System (1988–93) data adjusted for incompleteness of diagnostic coding using the following estimates of coverage by financial year: 1987–88 = 73%; 1989–90 = 85%; 1990–91 = 88%; 1991–92 = 96%; 1992–93 = 100%; 1993–94= 100%.

Scottish Hospital Inpatient Statistics (1982–87) plus Scottish Health Statistics (188–93).

Welsh data (1982–90 only) from Hyndman S, Williams DRR, Merrill SL, Lipscombe JM, Palmer CR. Rates of admission to hospital for asthma. BMJ 1994; 308: 159–1600.

Table 7A.5 **Hospital admission rates for asthma (per 10,000 per year), both sexes, by age, England and Wales 1962–85 and England only 1988–91**

Year	0–4	5–9	10–14	15–24	25–34	35–44	45–64	65–84	85+
					Age group				
England and Wales (HIPE)									
1962	4.03	4.35	3.20	2.99	3.74	3.26	5.14	3.92	0.69
1963	5.66	5.67	3.84	2.70	4.43	4.80	6.59	4.73	1.30
1964	5.74	6.50	5.58	3.66	4.74	5.51	7.33	4.34	1.27
1965	7.51	7.36	5.55	4.42	5.04	5.70	8.06	5.55	2.16
1966	6.71	8.52	6.95	4.44	5.01	5.76	7.95	5.80	1.86
1967	9.88	9.25	8.46	5.50	5.97	6.62	8.64	5.94	2.65
1968	9.94	11.29	8.86	5.40	6.64	6.10	8.79	6.82	4.09
1969	11.29	10.25	7.33	5.53	6.24	6.64	8.28	6.39	3.92
1970	11.02	10.78	6.01	4.76	4.46	5.27	7.42	5.75	1.69
1971	12.82	12.05	6.71	4.72	4.74	6.10	7.17	5.54	2.83
1972	11.20	10.09	7.41	4.92	4.12	5.29	7.57	6.24	5.12
1973	15.53	12.26	7.64	4.65	3.69	5.65	7.01	6.26	2.61
1974	13.53	10.74	6.21	4.52	3.87	4.90	6.59	6.62	3.12
1975	15.77	12.38	8.47	4.05	3.88	5.49	6.71	5.51	2.33
1976	19.15	13.89	7.62	5.20	3.67	5.60	7.17	6.15	3.38
1977	20.97	15.01	9.78	5.48	4.10	5.93	6.96	5.95	3.13
1978	26.04	17.49	9.66	5.84	4.54	5.59	6.97	6.65	2.67
1979	26.59	19.22	9.66	5.78	4.26	5.91	7.00	6.52	2.44
1980	34.01	22.02	11.73	5.88	4.59	6.11	7.42	7.18	4.93
1981	43.52	27.74	13.85	7.83	5.77	6.23	8.08	8.08	4.48
1982	53.21	31.37	15.26	7.95	5.49	6.21	8.80	8.80	4.55
1983	59.34	38.10	14.72	8.91	6.19	5.74	9.33	9.04	3.06
1984	71.80	39.31	16.67	8.88	6.61	6.63	9.94	10.55	6.31
1985	75.56	40.33	17.31	10.33	8.22	6.82	11.58	10.72	6.20
England (HES)									
1988	101.47	42.18	22.64	12.96	9.21	7.88	11.25	10.56	6.33
1989	93.10	36.06	23.29	13.58	9.47	8.16	11.72	11.15	6.54
1990	102.60	35.84	23.60	13.91	9.30	8.00	10.92	10.67	7.08
1991	94.38	35.79	23.27	14.61	9.95	8.16	10.52	11.11	7.63
1992	91.34	30.04	21.17	14.23	9.50	8.35	10.85	11.42	8.00
1993	93.61	30.31	23.05	15.27	10.76	9.40	11.80	13.36	9.52

Sources: Hospital Inpatient Enquiry (10% sample, 1962–85), Hospital Episode System (100% sample, 1988–93). HES data adjusted for incompleteness of diagnosis (see footnote to Table 7A.4).

Table 7A.6 **Mortality rates for asthma (per million per year) by age and sex, England and Wales, 1958–1992**

	Males					Females				
Year	**0–14**	**15–44**	**45–64**	**65–74**	**75+**	**0–14**	**15–44**	**45–64**	**65–74**	**75+**
1958	5.1	12.1	47.3	111.0	100.6	3.4	19.5	52.9	94.7	107.3
1959	4.3	7.2	44.3	92.1	106.0	3.9	14.5	42.6	75.5	91.4
1960	5.0	10.0	39.3	90.1	74.0	5.3	12.2	44.8	66.8	93.4
1961	4.8	10.6	39.8	103.5	94.6	6.8	15.7	39.2	80.3	92.3
1962	5.9	12.5	43.5	80.1	110.2	4.6	15.4	49.5	85.2	85.2
1963	9.5	16.4	50.6	88.2	98.6	6.7	24.6	63.9	81.5	97.6
1964	10.5	18.9	58.1	94.1	98.2	7.4	30.3	61.2	90.1	93.3
1965	12.8	27.2	60.0	104.1	95.4	10.0	32.7	78.0	93.8	90.9
1966	11.8	27.3	56.1	118.3	100.5	9.3	30.8	69.3	92.8	107.7
1967	12.3	21.4	52.3	84.7	90.1	8.4	25.2	58.1	90.0	86.8
1968	8.7	15.9	38.9	86.2	105.8	5.6	17.9	49.2	82.9	87.7
1969	7.8	12.6	38.4	91.5	67.9	3.7	12.6	44.6	78.0	73.1
1970	4.0	12.1	34.2	67.8	79.2	4.8	13.6	40.1	80.1	82.1
1971	4.8	12.9	31.1	57.4	69.4	5.1	12.8	45.0	69.0	82.1
1972	5.7	12.0	35.8	62.0	78.0	3.2	12.5	44.8	78.6	89.0
1973	5.8	9.3	33.0	74.4	63.0	3.3	10.9	40.8	80.3	85.9
1974	3.7	8.3	26.9	60.2	67.4	3.6	8.9	38.2	70.2	83.6
1975	3.9	10.4	32.4	68.0	101.3	2.5	9.2	37.5	73.6	91.9
1976	3.8	8.5	22.9	57.6	69.2	3.1	7.0	40.0	67.1	94.1
1977	4.8	9.2	22.9	52.5	74.2	4.7	8.1	36.7	72.2	137.2
1978	4.5	11.2	23.1	54.5	89.2	3.5	9.4	39.4	69.5	135.4
1979	3.3	10.2	34.8	80.8	122.3	3.7	11.5	41.1	100.2	116.3
1980	4.4	10.0	40.2	84.3	102.0	3.4	8.9	43.9	95.4	130.3
1981	4.6	13.7	43.3	85.2	122.0	4.9	10.2	54.3	94.6	126.5
1982	6.7	12.3	36.6	81.9	140.7	2.1	9.9	50.5	98.5	159.9
1983	4.8	12.2	45.7	88.9	127.2	3.8	10.0	48.7	111.8	184.9
1984	5.3	13.4	44.0	103.1	167.0	2.8	8.9	47.4	108.5	175.5
1985	4.1	12.7	51.8	101.1	177.0	2.6	13.0	58.3	106.6	180.5
1986	6.2	13.2	50.4	110.1	162.1	2.8	11.6	57.6	126.9	188.2
1987	6.2	13.9	53.2	84.8	153.2	4.2	11.2	46.9	115.5	180.5
1988	3.9	12.3	45.0	107.5	188.1	4.2	11.4	53.9	121.0	188.2
1989	3.1	12.2	39.9	94.3	189.1	2.2	9.5	49.0	132.4	197.2
1990	5.7	9.7	35.6	105.5	161.3	1.9	9.4	47.9	109.8	199.4
1991	4.0	11.1	46.6	94.7	179.0	2.5	8.6	39.8	117.0	190.7
1992	2.4	9.0	36.6	80.5	188.0	1.7	7.0	38.2	111.7	208.7

Source: Historical mortality file, OPCS.

Table 7A.7 **Mean and peak smoke and sulphur dioxide concentrations measured during October to March 1962–63 in cities and county boroughs of England, Wales and Scotland**

County borough	Winter mean		Winter peak		Grid ref		No of
	Smoke	SO₂	Smoke	SO₂	E	N	sites
Wigan	323	143	2	0	358	405	1
Birkenhead	206	295	2	3	332	388	3
Chester	254	182	2	2	340	366	2
Stockport	352	327	2	3	389	390	3
Wallasey	259	323	2	3	329	392	3
Barrow	224	207	1	1	319	469	1
Blackburn	391	304	3	3	368	428	3
Blackpool	297	246	2	2	330	435	3
Bolton	263	324	2	3	371	408	3
Bootle	460	479	3	3	333	394	1
Burnley	237	301	2	3	384	432	3
Bury	334	362	2	3	380	410	2
Liverpool	175	500	0	2	334	390	1
Manchester	314	338	3	3	383	397	4
Oldham	369	294	2	3	392	404	4
Preston	396	380	2	2	354	429	1
Warrington	374	378	2	3	360	388	3
St Helens	179	371	2	3	350	395	2
Salford	494	556	3	3	380	398	3
Darlington	225	—	1	—	429	514	2
Gateshead	245	216	2	1	425	560	1
S Shields	337	216	2	2	436	566	1
Sunderland	311	207	2	2	439	557	5
Hartlepool	420	279	2	1	451	532	1
Newcastle	330	280	3	3	424	564	5
Tynemouth	320	121	2	1	436	569	1
Teesside	407	249	2	2	449	520	3
K-u-Hull	226	216	2	1	509	428	1
Bradford	309	341	3	3	416	433	8
Doncaster	299	254	2	2	457	402	3
Halifax	229	313	2	3	409	425	4
Huddersfield	344	362	3	3	414	416	5
Leeds	382	438	3	3	430	434	6
Sheffield	287	277	3	3	435	387	6
Wakefield	339	330	3	3	433	420	3
York	232	213	2	2	460	452	2
Derby	365	342	3	3	435	336	2
Leicester	222	247	2	2	458	304	5
Lincoln	212	137	1	1	497	371	4
Nottingham	234	203	3	3	457	341	5
Norwich	172	134	1	1	623	308	3
Ispwich	185	166	2	2	616	244	3
Southend	184	172	2	2	587	185	1
Cen London	204	371	3	3	532	180	9
NE London	169	255	2	3	545	186	5
SE London	134	215	2	3	540	169	3
NW London	161	278	2	3	520	184	9
SW London	149	188	2	3	525	169	7
Newham	169	255	2	3	542	182	5
Croydon	188	282	3	3	533	165	3
Eastbourne	119	111	0	0	561	99	3
Reading	121	160	1	2	471	173	3
Oxford	152	195	2	1	451	206	3
Portsmouth	88	99	0	0	465	101	3
Southampton	136	135	1	2	441	112	2
Exeter	93	70	0	1	292	92	2
Plymouth	167	99	1	0	247	55	1
Cardiff	98	127	0	0	318	176	4
Swansea	70	83	0	2	265	193	4

Table 7A.7 continued

County borough	Winter mean		Winter peak		Grid ref		No of
	Smoke	SO$_2$	Smoke	SO$_2$	E	N	sites
Newport	88	158	2	2	331	188	4
Warley	244	369	2	2	402	288	2
Stoke	340	420	2	3	387	345	1
Walsall	330	400	3	3	401	298	2
W Bromwich	232	305	2	1	400	292	4
Birmingham	277	400	2	1	407	297	2
Coventry	165	220	2	1	433	279	3
Dudley	204	179	2	1	394	290	3
Motherwell	210	148	2	1	275	657	7
Paisley	353	179	3	2	243	663	2
Dumbarton	261	100	3	1	255	674	2
Aberdeen	218	116	2	0	394	806	2
Dundee	230	131	1	0	340	730	2
Edinburgh	191	111	1	0	325	673	6
Glasgow	352	235	3	2	258	665	7

Sources: National Survey of Smoke and Sulphur Dioxide 1962–63

Mean concentrations calculated as the average of published figures for the number of air pollution monitoring sites listed. (Mean figures were not published for sites with substantial amounts of missing data.)

Peak concentrations:	$0 = \leqslant 500\,\mu g/m^3$ at one or more sites
(highest daily mean)	$1 = 500–999\,\mu g/m^3$ at one or more sites
	$2 = 1000–1499\,\mu g/m^3$ at one or more sites
	$3 = > 1500\,\mu g/m^3$ at one or more sites

127

Table 7A.8 **Average annual mean smoke and SO₂ levels ($\mu g/m^3$) in 1991 at non-industrial sites by county or Scottish region**

Area	SO₂	Smoke
Metropolitan English counties		
Greater London	21	17
Greater Manchester	37	14
Merseyside	39	13
South Yorkshire	56	24
Tyne & Wear	33	22
West Midlands	34	14
West Yorkshire	43	17
Non-metropolitan English counties		
Avon	no data	no data
Bedfordshire	no data	no data
Berkshire	40	6
Buckinghamshire	no data	no data
Cambridgeshire	no data	no data
Cheshire	31	12
Cleveland	27	9
Cornwall	no data	no data
Cumbria	no data	no data
Derbyshire	48	17
Devon	no data	no data
Dorset	no data	no data
Durham	36	24
East Sussex	48	7
Essex	no data	no data
Gloucestershire	no data	no data
Hampshire	no data	no data
Hereford & Worcestershire	no data	no data
Hertfordshire	no data	no data
Humberside	18	8
Isle of Wight	no data	no data
Kent	28	13
Lancashire	no data	no data
Leicestershire	no data	no data
Lincolnshire	35	10
Norfolk	16	12
Northamptonshire	27	18
Northumberland	no data	no data
North Yorkshire	20	11
Nottinghamshire	46	19
Oxfordshire	18	18
Shropshire	no data	no data
Somerset	no data	no data
Staffordshire	63	27
Suffolk	no data	no data
Surrey	22	11
Warwickshire	no data	no data
West Sussex	no data	no data
Wiltshire	no data	no data
Welsh counties		
Clwyd	no data	no data
Dyfed	12	7
Gwent	no data	no data
Gwynedd	no data	no data
Mid Glamorgan	no data	no data
South Glamorgan	no data	no data
West Glamorgan	44	7
Scottish regions		
Highland	no data	no data
Grampian	19	3

Table 7A.8 continued

Area	SO$_2$	Smoke
Tayside	16	9
Fife	no data	no data
Lothian	27	16
Central	21	8
Borders	no data	no data
Strathclyde	39	24
Dumfries & Galloway	no data	no data
Orkney	no data	no data
Shetland	no data	no data
Western Isles	no data	no data

Source: Warren Spring Laboratory.

Purely industrial excluded. The average of annual means recorded for other sites in each county in 1991 is tabulated.

Chapter 8

Air Pollution and the Timing of Asthma Attacks: Population Studies

Introduction

8.1   This chapter reviews the evidence available from studies of *whole populations* concerning the relationship of short-term fluctuations in the incidence of asthma attacks to changes in air pollution levels. This will include consideration of "asthma epidemic days", historical air pollution episodes and analyses of extended time-series in the UK and elsewhere. Experimental and observational studies of the acute effects of air pollutants on *individuals* are reviewed in Chapter 5 and 6, respectively.

8.2   Short-term fluctuations in asthma occurrence, along with the results of chamber and panel studies (Chapters 5 and 6), contribute to our knowledge about factors which may *exacerbate* asthma. They are less relevant to determining whether the *prevalence* of asthma is related to current or past air pollution exposure.

Asthma epidemic days

8.3   Although there is a marked seasonal variation in the incidence of asthma attacks, there have been relatively few reports of noticeable "epidemics" of asthmatic episodes lasting for a day or two at a time. Relatively small numbers of daily attendances due to asthma can be expected at a single hospital, health centre or emergency department, and chance fluctuations may obscure all but the most dramatic clusters of asthma attacks. Retrospective exploration of daily rates of health service utilisation has identified epidemic days which were not clinically apparent at the time.[1] Thus, it is possible that reported epidemics are the tip of an iceberg of daily variation due to unrecognised "unusual" days.

New Orleans

8.4   Asthma epidemic days were a recognised phenomenon in New Orleans during the 1950s and 1960s.[1] On certain days during the autumn, and occasionally in the spring, the number of emergency room attendances at a major hospital rose to 100–250, compared to an expected rate of 15–25 per day. The epidemics apparently began in 1953 and typically occurred during the first three weeks of October each year, the driest month in New Orleans. Telephone surveys established that on the same days, asthmatic patients who did not attend the hospital also experienced attacks, and the epidemic appeared to affect much of the city. Air pollution, possibly related to burning of rubbish dumps, was suspected initially, but the epidemics continued despite cessation of this seasonal activity.[1] No temporal correlation was found between asthma attendances and gaseous pollutants (SO_2, NO_2, or aldehydes), but levels of small airborne particles were higher on epidemic days.[2] These particles included pollens and spores. The individuals involved in epidemics were predominantly atopic asthmatic patients, with positive skin prick tests to many common aeroallergens.[3] Positive reactions to grain dust were present in 80% of the affected cases, but the prevailing wind and pattern of operation of a nearby grain elevator did not suggest this as a likely cause of the epidemic days. Detailed investigations of airborne pollen and fungal spores failed to identify a specific allergen that might be responsible, but it was noted that autumn epidemic days tended to be associated with high total spore and pollen counts.[3] The cause of New Orleans epidemics remains uncertain, but it is considered more likely to be aeroallergens than non-biological air pollution.

Barcelona

8.5   Eight outbreaks of asthma were recorded in Barcelona between 1981 and 1986.[4] These were identified by an increase in daily hospital attendances at adult emergency rooms, although no increase was apparent at paediatric hospitals. The admissions clustered around midday and were characterised by a rapid onset of severe asthma requiring intensive care. In early epidemics, the spatial pattern of attacks could not be described in detail because of inadequate data from many hospitals, but in the eighth outbreak (November 1986) complete admission data was obtained, and a marked clustering of attacks was noted in a central area of the city close to the docks. Compared to other days in the surrounding month, asthma admissions on 26 November were increased by a factor of three overall, and to a greater relative extent in the elderly (ninefold increase) than in children (2.5-fold increase). No increase in admissions for other respiratory diseases occurred. Air pollution levels and airborne pollen and spore counts on the epidemic day were below average for the city. Subsequent investigation implicated the unloading of soybeans in the harbour as the cause of this and other epidemics in Barcelona.[5] Patients affected on epidemic days were found to be much more likely to be allergic to soybean, as measured by levels of serum IgE specific to soybean.[6] The prime cause of the Barcelona epidemics was thus firmly established as aeroallergen pollution associated with industrial activity at the dockside, and no further epidemics have occurred since modifications were made to the procedures for unloading soybeans.

8.6   At first, air pollution was considered as a possible cause of the Barcelona epidemics,[7] but was subsequently discounted because pollution levels were not unusually high at the time of the epidemics and patients who had been admitted during epidemics did not report worsening of symptoms at other times in association with increases in NO_2 and SO_2 pollution.[8] Although aeroallergens have been implicated as the necessary cause, the possibility of a biological interaction between allergen and pollutants in Barcelona has been raised by a recent abstract.[9] During 1985–87, soybeans were unloaded on 123 days, of which only 13 were associated with asthma epidemics. The levels of NO_2 and SO_2 were somewhat higher on these 13 days than on the other 110, after adjustment for climatic conditions, suggesting that air pollution may have played a synergistic role in the Barcelona epidemics caused by population exposure to soybean dust.

Birmingham, UK.

8.7   An epidemic of asthma occurred in central England in July 1983, associated with a severe thunderstorm on the evening of 6 July.[10] Over a 48-hour period, 106 patients attended accident and emergency departments of hospitals in Birmingham, compared to an expected rate of about 10 per day. There was also evidence of an excess of hospital attendances and general practitioner consultations for asthma throughout central England. Many of the patients had a history of hay fever and seasonal asthma, suggesting an atopic disposition. Airborne smoke and SO_2 levels were not increased at the time of the outbreak (51 $\mu g/m^3$ and 111 $\mu g/m^3$ (39 ppb), respectively). Airborne spore counts of the moulds *Didymella exitialis* and *Sporobolomyces* rose markedly during or shortly after the thunderstorm and some patients with late summer asthma may be allergic to *Didymella* spores.[11] However, it remains unclear why that thunderstorm, and not others, should have resulted in such a marked increase in asthma attacks.

Melbourne

8.8   Three epidemics of asthma in association with thunderstorms have been reported from Melbourne.[12,13] Each occurred in November (early summer) in the years 1984, 1987 and 1989, suggesting that some seasonal factor in addition to the thunderstorm might be implicated. Levels of ozone, NO_2 and SO_2 were low and no higher than usual on epidemic days. Patients admitted during the epidemic all suffered from hay fever, were allergic to rye grass pollen and had significantly larger skin prick reactions to this allergen than asthmatic patients not affected by the storms.[13] One of the potent allergens of rye grass pollen (*Lol pIX*) has been identified in starch granules contained within pollen grains.[14] These granules are released in large numbers when the pollen comes into contact with water, as might occur during a thunderstorm, and are small enough to enter the airways (whereas intact pollen impacts in the nose). Four patients with thunderstorm-associated asthma had marked bronchoconstriction after experimental inhalation of these starch granules.[14]

This suggests that allergic reactions to osmotically disrupted rye grass pollen were probably the cause of the Melbourne outbreaks, and may have been a factor also in the epidemic associated with a thunderstorm in central England.

London and southeast England

8.9 An asthma epidemic of unprecedented proportions occurred on the night of 24–25 June 1994 in association with thunderstorms across southeast England.[15] Preliminary data suggest a ten-fold increase in accident and emergency attendances for asthma in the worst affected areas (northeast London) with a three-fold increase in admissions for asthma on the night of the storm. Emergency calls to general practitioners were also increased.[16] The vast majority of affected patients were hay fever sufferers, although many had not previously been diagnosed asthmatic. The storm followed two days of high grass pollen levels, but air pollutant concentrations were unexceptional for the time of year. The circumstantial evidence thus suggests that disrupted grass pollen was probably the cause of this most recent epidemic, and studies are underway to investigate this.

Comment

8.10 The patients affected by defined epidemics of asthma appear to be atopic asthmatics and although the precise nature of the allergen is uncertain in New Orleans and central England, aeroallergens have been implicated as a cause of each of the four major series of asthma epidemic days. Geographical clusters of asthma have also been reported in association with allergen pollution by castor-bean dust.[17–19] Gaseous air pollution appears to play little, if any, role in recognised outbreaks of asthma, and the role of non-biological particulate pollution is difficult to define because all of the allergens which have been implicated are present in the particulate fraction.

Air pollution episodes

London, 1952

8.11 The most notorious episode of air pollution occurred in London during December 1952, when cold, still weather conditions led to the formation of a temperature inversion over the Thames valley and failure of the normal atmospheric dispersion of smoke and SO_2 pollution from local industry and domestic coal fires. Daily average SO_2 concentrations in central London were $3,830\,\mu g/m^3$ (1,339 ppb) on the worst days, and the maximum daily smoke levels at central sites could not be estimated because the samplers were overloaded: a conservative estimate being $4,460\,\mu g/m^3$.[20] These values were at least ten times the daily average for a typical winter in London in those years, and about 100 times the daily urban values nowadays. Lower absolute values, but similar relative increases in pollutant levels were measured in outer London.

8.12 Evaluation of the health effects of the London smog concentrated on mortality, but some consideration was given to hospital admissions and consultations in general practice. As in subsequent investigations in London,[21] the focus was primarily on chronic bronchitis rather than asthma. Only ten deaths occurred from asthma during the period of the smog, compared to an expected number of two. Out of 44 deaths which were ascribed to the fog, 6 had a definite history of asthma and 9 possibly suffered from the disease. However, the cause of death in these patients was not necessarily due to bronchospasm.[20]

8.13 Observations made in general practice in the outer London suburb of Beckenham (where visibility, and therefore presumably pollution levels, were much less affected than in central London) suggested that the majority (37/43) of the patients consulting with lower respiratory disorders during the 1952 smog were already known to have asthma, bronchitis or another chronic lung disease.[22] Only six of these were under 50 years of age, and typically their illness commenced suddenly on the third day of the fog. The clinical picture was a combination of mucosal irritation (cough and sputum), bronchial obstruction (breathlessness and audible wheezing), infection (fever and generalised upset) and cardiac failure in the elderly.

The wheezy breathlessness did not respond to the usual antispasmodic treatment in use at the time (adrenaline, ephedrine, isoprenaline and aminophylline). None of the asthmatic children in the practice were affected by the fog. A very similar set of clinical problems, again sparing children, was presented during a less severe episode of smoke and SO_2 pollution in London in December 1962.[23]

Meuse valley, 1930 and Donora, Pennsylvania, 1948

8.14 In earlier episodes of severe smoke and SO_2 pollution in the Meuse valley[24] and Donora, Pennsylvania,[25] smaller populations were affected but there was more detailed investigation of individual fatalities. Asthma was considered to have played a part in 5 of 12 fatalities investigated in Donora, and most of the 50 people hospitalised during that episode were previously diagnosed asthmatic patients. In the Meuse episode, respiratory symptoms associated with the smog were partially relieved by subcutaneous adrenaline, but the autopsy findings suggested that the cause of the 60 deaths was often pulmonary oedema, a consequence of acute heart failure rather than asthma. This raised the possibility that sudden deaths ascribed to "bronchial asthma" may have been related to "cardiac asthma" due to pulmonary oedema resulting from heart failure.

New York, 1953 and 1962

8.15 Well defined episodes of smoke and SO_2 pollution occurred in New York periodically throughout the 1950s and 1960s. An unusually severe episode in November in 1953 was associated with increased all-cause mortality and an increase in paediatric and adult clinic attendances at four major hospitals during the period of the smog.[26] Clinic visits for asthma were specifically investigated and no increase was evident during or after the period of maximal pollution. No excess of asthma attendances occurred at five New York hospitals during a second severe episode in December 1962.[27]

New York, 1966

8.16 A further episode of smog associated with a temperature inversion and anticyclonic weather affected New York for three days over Thanksgiving Day weekend in November 1966.[28] Temperature was not unusually low for the time of year. Sulphur dioxide levels during the episode were approximately 1,450 μg/m^3 (507 ppb) (daily averages), about three times the usual level for November during 1961–65. Peaks of up to 2,900 μg/m^3 (1,014 ppb) SO_2 (maximum hourly average) were recorded on the third day. There was also a threefold increase in levels of particles, as measured by the (now obsolete) smoke shade technique. Attendances at emergency clinics at seven New York hospitals due to asthma and bronchitis (combined) were increased only on the third day of the episode, and this excess was only apparent in the over-45 age-group, among whom other forms of chronic obstructive airways disease are more prevalent than asthma. Interpretation of these findings is severely complicated by the closure of most outpatient clinics on the Thanksgiving Day holiday on the second day of the smog, and the absence of data from previous years with which to estimate the "rebound" in daily attendances after a public holiday.

8.17 A more detailed analysis of children who had visited the emergency room of one New York hospital during November and December 1966 revealed a statistically significant excess of attendances with respiratory symptoms in the week of the smog.[29] This was largely attributable to a remarkable cluster of 15 patients with "symptoms of expiratory obstruction" attending on the two days immediately *after* the three days of highest pollution levels, compared to an average of about one per day during the remainder of the period (including the smog days). This suggested a 72-hour lag between exposure to pollution and presentation to hospital, similar to the experience in general practice in the London smogs, but detailed information on the time of onset of symptoms was not obtained. A collation of admission data from other paediatric hospitals did not reveal an obvious excess during the week of the 1966 smog.[28]

8.18   Within 48 hours of the end of the November 1966 smog, 2,052 employees of a New York insurance company were asked to complete a questionnaire on respiratory symptoms experienced during the air pollution episode.[30] The period of recall included three days during and two days after the smog. Unfortunately, the occurrence of symptoms in the few days preceding the pollution episode was not ascertained. Thus, there is no baseline data with which delayed effects of the smog may be compared. The reported prevalence of most symptoms increased from day 1 to day 3 (smog period) and declined thereafter. There was a modest rise in the prevalence of wheeze and difficulty in breathing (a relative increase of about 50% in each) on day 3 compared to days 1 or 5.

Mount St Helens eruption, 1980

8.19   Exceptional levels of particulate pollution were experienced by residents of Yakima, Washington state, USA, following three volcanic eruptions of Mount St Helens during May and June 1980.[31] During the first eruption, falling ash caused almost total darkness at times and total suspended particles exceeded the capacity of the measuring instrument ($>30,000 \mu g/m^3$), remaining above $1,000 \mu g/m^3$ for 8 days. Levels of gaseous air pollutants are not reported. The number of emergency room visits for asthma in two major local hospitals was noticeably elevated for 7 of these 8 days (10 or more per day, compared to a usual average of 2–3 per day). Ninety per cent of patients who attended the emergency rooms with asthma over the four weeks of the eruptions were already known to be asthmatic subjects. Interviews with other known asthmatic patients established that the dustfall had also exacerbated symptoms in patients who did not present to hospital.

The Ruhr, Germany, 1985

8.20   In January 1985, the Ruhr area of West Germany experienced a smog episode for 5 days, associated with a temperature inversion and a slow easterly wind which imported pollutants from Czechoslovakia and East Germany.[32] Sulphur dioxide levels reached $830 \mu g/m^3$ (290 ppb) (maximum 30-minute average), and suspended particles were also markedly increased (maximum daily average $600 \mu g/m^3$). Nitrogen dioxide levels were increased to a lesser extent (maximum daily average $230 \mu g/m^3$ (122 ppb)). Mortality, hospital admissions, ambulance journeys and consultations in primary care were compared before, during and after the smog for the polluted area and a control area where pollution levels, though increased above their usual levels, were lower than in the smog zone (maximum daily averages: $320 \mu g/m^3$ (112 ppb) SO_2; $120 \mu g/m^3$ (64 ppb) NO_2; and $190 \mu g/m^3$ particles). Data specific to asthma are presented only for hospital admissions. There was a non-significant *decrease* by 14% in the asthma admission rate during the smog period in the polluted area, and a 4% decrease during the smog period in the control area. During the smog period in the polluted area there was a non-significant 12% increase in admissions for obstructive bronchitis (which probably includes bronchiolitis as diagnosed among infants in Britain). No comparable data on obstructive bronchitis are presented for the control area, so it is not possible to assess the effect of the annual winter epidemic of respiratory syncytial virus infection, which is closely and causally related to bronchiolitis admissions.

Central England, 1985

8.21   The same weather conditions which caused smog in West Germany also resulted in a marked increase in SO_2 concentrations over central England.[33] The mean daily SO_2 level recorded in the week of the episode was $104 \mu g/m^3$ (36 ppb) in the polluted area and $72 \mu g/m^3$ (25 ppb) in the rest of the country. There was no associated rise in levels of airborne particles. During the episode week, no rise in consultations was evident for either acute asthma or acute bronchitis among the "spotter" general practices in the polluted area (serving over 100,000 people). There was no excess of consultations for asthma or bronchitis in the polluted area by comparison with the rest of the country. This was a relatively mild pollution event and the health effects were measured on a weekly, rather than a daily basis, which may have obscured subtle variations in asthma incidence.

8.22   An episode of exceptional NO_2 pollution occurred in London in December 1991, when hourly average NO_2 levels reached 423 ppb (795 $\mu g/m^3$), well in excess of the WHO guideline (395 $\mu g/m^3$ (210 ppb)) and higher than any previous measurement at a non-kerbside site anywhere in the UK. Sulphur dioxide levels were not increased, and ozone levels were low, but Black Smoke samplers registered levels of up to 228 $\mu g/m^3$ (daily average).[34] Comparison of admission rates in Greater London and the surrounding south-east of England during the episode week with those in the preceding week and in previous years showed a small and non-significant 3% increase in asthma admissions from the level predicted in the capital during the episode week. Other respiratory outcomes, notably chronic bronchitis and emphysema among the elderly, were affected to a greater extent than asthma admissions. This may be partially attributable to the rise in particulate pollution, rather than to NO_2 toxicity.[34]

Comment

8.23   There is little doubt that early smog episodes resulted in excess mortality, but the nature of this acute toxicity and its relationship, if any, to asthma is difficult to evaluate from the published reports. The clinical experience of non-fatal respiratory complications in the more severe smogs appears to have been exacerbation of existing chronic respiratory disease in middle-aged and older adults with a latent period of 2–3 days. This suggests as a mechanism secondary infection rather than acute bronchospasm. There is little indication of an effect of these smogs on asthmatic children and young adults. In more recent urban smogs with lower levels of SO_2 (New York 1966 and The Ruhr 1985), there is no evidence for a general increase in asthma attack rate, except in one New York children's hospital in 1966. However, in the exceptional circumstances of the Mount St Helens eruption, extremely high levels of particles (in association with unknown concentrations of volcanic gases) did increase the incidence of asthma substantially and with a short latent period.

8.24   Only the most recent London NO_2 episode is relevant to evaluating the health effects of "new" pollutants, and even here there was a modest increase in levels of particles. It should be noted that the peak hourly levels of ambient NO_2 recorded outdoors during this episode would not be considered unusual in a kitchen where gas was used for cooking.[35] These levels of exposure were not associated with a substantial or significant increase in hospital admissions for asthma in Greater London, despite considerable concern and publicity about this at the time.

Time series analyses   8.25   This section reviews evidence from analyses of hospital admissions or emergency health service contacts for asthma in relation to short-term (usually daily) fluctuations in air pollution levels. The time series method has been applied to studies of mortality, hospital admissions, emergency room visits and school absence. None of the studies of school absence have published results specifically for asthma,[36-38] and death from asthma is too rare an event to be analysed on a daily basis. A number of studies of health service contacts for respiratory diseases were excluded from consideration here because they did not report findings separately for asthma or wheezing illness,[39-45] or because they specifically excluded asthma.[46]

Methodological issues   8.26   Simple correlations of daily or weekly asthma attack rates and air pollution levels are prone to a number of statistical artefacts which may tend to exaggerate the association between air pollution levels and the day-to-day variations in asthma incidence.[45,47] Sophisticated statistical methods have been developed to overcome some of these problems, as discussed below. These usually require the derivation of a *regression model* which relates the disease outcome to a number of explanatory variables (season, day of the week, temperature, pollutant levels etc) in the form of a mathematical equation.

Serial correlation (autocorrelation)

8.27   The most serious problem complicating the analysis of time series is the tendency for both the outcome (asthma incidence) and the explanatory variables (air pollution and climatic variables) to exhibit cyclical fluctuations throughout the year.

These long-wave cycles may generate a spuriously high degree of correlation between daily measures, or (if they are out of phase) may obscure correlations over a shorter time scale. Two methods have, in general, been used to filter out this long wave autocorrelation and concentrate the analysis on short-term variations over a period of less than one month: subtraction of a moving average from each daily count of asthma occurrence; or introduction of model terms describing long-wave cycles (of varying periodicity) into the regression equation. In theory, short-wave autocorrelation may remain after such adjustments, although in practice this does not seem to be a major problem. The more advanced analyses use *autoregressive* modelling which takes account of serial correlation of the observations from adjacent days. This tends to produce more conservative tests of statistical significance if autocorrelation is present.

Small numbers

8.28 Even in a population centre of moderate size, the number of asthma admissions or health service contacts per day is often small (less than 10 per day). In such circumstances it is inappropriate to assume a statistically normal (Gaussian) distribution for chance fluctuations from the underlying pattern of asthma incidence. Regression models with *Poisson errors* are more suitable and assign less weight to days with zero counts than would a conventional regression or correlation analysis with Gaussian errors. On the other hand, there are theoretical arguments that asthma incidence may vary from day to day for reasons other than air pollution or Poisson sampling variability. It is possible to evaluate such *extra-Poisson variation*, although few analyses have done so. Failure to account for such additional variability, if it were present, would result in too great a weight being assigned to days with high asthma counts.

Use of indirect indicators of disease

8.29 Use of hospital and other health services for asthma may vary over time for reasons unrelated to the incidence of asthma attacks. The most important of these relate to weekends, holiday periods and accessibility, particularly during severe weather. The regression model may be improved by including dummy variables for these factors. Admission rates may also be influenced by the availability of beds (and thereby by exacerbations of other diseases), although this has not been addressed in any of the studies to date.

Measurement of pollution exposure

8.30 Most analyses rely on pollutant levels from a small number of sites (often only one site). The assumption is thereby made that fluctuations in the levels of pollutants at the monitoring site reflect short-term variations in pollutant levels over a wider area. Whereas this may be a reasonable simplification for daily concentrations of regionally dispersed pollutants such as ozone and airborne sulphates, more local monitoring may be required for a valid indication of fluctuations in particulate and industrial pollution. Peaks of exposure (measured as hourly maxima) tend to correlate less well than daily means over a wide area, but may be more relevant to the occurrence of health effects. A further uncertainty, which has rarely been addressed, is the extent to which short-term fluctuations in ambient exposure reflect variations in personal exposure, particularly in winter, when most of the population spend their time indoors. Imprecision in the assessment of individual exposure will tend to dilute any underlying association between pollutant exposure and disease.

Lag periods

8.31 There may be a delay between relevant pollutant exposures and ascertainment of a health effect, either because of a biological latent period, or due to delays in referral or presentation to hospital. Many analyses have, therefore, considered correlations between daily asthma attack rates and pollutant levels measured a few days before (lags 1, 2, 3 days etc). While this may be biologically appropriate, it increases the opportunity for multiple comparisons, and thereby of attaining a statistically significant result.

8.32   The more statistical associations that are examined in a set of data, the greater the number of "statistically significant" results obtained. There is ample opportunity for performing multiple comparisons in time series analyses, due to the inclusion of several pollutants; a choice of lag periods; and often several outcome measures, among which asthma may be only one. In addition, several investigators have examined subsections of their data, defined by age; season; year of study; or area of residence. Often the results quoted in most detail are those which are the most statistically significant. These require cautious interpretation as they are likely to exaggerate the overall relationship between pollution and disease.

Intercorrelation (colinearity) of pollutant levels

8.33   In many time series, several pollutants display similar short-term variability within a single area. Simultaneous modelling of the effects of highly correlated pollutants could lead to grossly misleading estimates of their "independent" effects. It is appropriate in such circumstances to include each pollutant singly in the regression model, acknowledging that it indicates the presence of a "pollution cocktail", rather than a specific causal agent. The corollary is that a single study can rarely determine the independent effect of any given pollutant with any certainty. Comparisons with similar analyses in other areas with different pollutant mixes may prove more informative.

Table 8.1   **Estimates of the relative change in asthma incidence for a 10 $\mu g/m^3$ increase in levels of individual pollutants as derived from published time-series analyses.**

Author(s)	Area of study	Period	Days	Events	Number per day	Pollutant and lag period	Relative Risk per 10 $\mu g/m^3$ ↑ (95% CI)	Comments
Walters et al[48]	Birmingham, UK	1988–90	730	Admissions	5	BS same day	0.98 (0.94–1.02)	Data for spring,
						BS lag 2d	1.00 (0.96–1.05)	summer and autumn
						BS 3d mean	0.95 (0.90–1.00)	pooled[a]
						SO₂ same day	1.00 (0.97–1.03)	Data for spring,
						SO₂ lag 2d	1.00 (0.97–1.03)	summer and autumn
						SO₂ 3d mean	1.00 (0.97–1.03)	pooled[a]
						BS same day	1.06 (0.97–1.14)	Data for winter:
						BS lag 2d	1.10 (1.01–1.19)	estimates may be
						BS 3d mean	1.06 (0.96–1.16)	inflated[b]
						SO₂ same day	1.05 (0.98–1.12)	Data for winter:
						SO₂ lag 2d	1.07 (1.00–1.15)	estimates may be
						SO₂ 3d mean	1.08 (1.00–1.16)	inflated[b]
Schwartz et al[49]	Seattle, USA	1989–90	395	ER visits	7	PM₁₀ 4d mean	1.04 (1.01–1.06)	Ages 0–65 only
						SO₂ same day	0.97 (0.81–1.19)	Ages 0–65 only
Pope[50]	Utah, USA	1985–89	1825	Admissions	c	PM₁₀ 1y mean	1.04 (includes 1)	See note[c]
Schwartz et al[49]	Seattle, USA	1989–90	153	ER visits	7	O₃ same day	0.99 (0.96–1.02)	5 summer months
Cody et al[51]	New Jersey, USA	1988–89	246	ER visits	3.6	O₃ lag 1 day	1.01 (1.00–1.02)	May to August
Thurston et al[52]	New York, USA	1988	92	Admissions	49	O₃ lag 1 day	1.01 (1.00–1.02)	June, July and
						SO₄ same day	1.07 (1.02–1.13)	August
						H⁺ same day	1.27 (1.09–1.44)	See note[d]

[a]   Published estimated for each season pooled by weighting inversely to their variance. These results are unlikely to have been influenced by the 1989 influenza epidemic, but exclude periods of high smoke (BS) levels

[b]   Relative risks likely to be spuriously raised (to an unknown extent) by the 1989 influenza epidemic

[c]   Estimates derived from a comparison of asthma admission rates in Utah and Salt Lake Valleys during a 13–month period of steel mill closure in Utah Valley, compared to other periods. There were about 0.5 admissions per day in Utah Valley and about 1 per day in Salt Lake Valley. The mill closure is presumed to have halved the annual mean PM₁₀ level in Utah Valley (from 53 $\mu g/m^3$ to 27 $\mu g/m^3$). This is the same relative reduction as quoted for the winter mean in the published report

[d]   For estimates of the relative influence of each pollutant across the range of level observed, see text

Conversion factors:	BS (Black Smoke) ≈ PM₁₀	O₃ (ozone) 1 $\mu g/m^3$ = 0.5 ppb
	SO₂ 1 $\mu g/m^3$ = 0.35 ppb	H⁺ (acid) measured as sulphuric acid equivalents

Measures of association

8.34   The correlation coefficient is the most widely used measure of association in time series studies. This is unfortunate as it conveys no quantitative information about the dose-response relationship between pollutant exposure and disease (only its direction). More recent analyses, particularly those with more complex and rigorous methodology, have quoted coefficients from regression models which permit an estimation of the dose-response gradient.[48-52] These are summarised in Table 8.1 which is discussed in the concluding section. In the following two sections, each individual study is summarised in one of two broad categories: studies primarily related to wintertime or to summertime pollution mixes.

"Winter smog"
pollution

Brisbane, Australia

8.35   An early study of the acute effects of smoke and SO_2 pollution on asthma incidence was based on nightly attendances for asthma at the Royal Brisbane Hospital during the winters of 1960–62.[53] On average, there was only one attendance per night, so the weekly number of attendances was analysed. A simple adjustment for seasonality was made by comparing the observed attendances with the number expected from the seasonal variation in temperature. No correlation was found between levels of particles (measured as coefficient of haze) and asthma attendances. An increased number of night visits for asthma occurred on only one of the 18 most polluted days during the study period. This was a small study and a simple analysis which may have had insufficient power to detect subtle effects. However, it excluded a major influence of wintertime particulate pollution in the timing of asthma attack in Brisbane.

Philadelphia, USA

8.36   Emergency room visits for asthma to a children's hospital in Philadelphia were studied over the period July 1963 to May 1965.[54] The hospital was located in an area with high dust levels, but no pollutant data are presented. Visits increased by up to ninefold on stagnant days which were reported to be associated with high levels of pollution. Asthma incidence did not correlate with pollen counts or the occurrence of epidemics of upper respiratory infection. These are unusual findings which are difficult to interpret in the absence of specific pollutant measurements.

New York, USA

8.37   Early studies of the relationship between levels of SO_2 and particles in New York and emergency room visits for asthma show no evidence of any relationship to particulate pollution (measured as coefficient of haze) or SO_2 levels[55] or day-to-day changes in pollution.[56] Each of these analyses was based on large numbers of emergency visits and took care to adjust for seasonal variations in asthma incidence by comparing each day's attendances either with the previous day[56] or the surrounding fortnight.[55] However, as they were based on daily average SO_2 concentrations, they could have missed the acute effects of short-term peaks of SO_2 exposure, similar in duration to those found to affect asthmatic patients in chamber experiments.

8.38   Such short-term effects were addressed specifically in an analysis of asthma presenting to hospital emergency rooms in Harlem (more polluted) and Brooklyn (less polluted) during 1969–72.[57] There were about 11 asthma attendances per day in Harlem and 22 per day in Brooklyn. Seasonal cycles were removed by comparing the number of daily attendances with the average daily number in the preceding and subsequent week. Each day was classified as less than or greater than the expected number, and days with unusually high asthma incidence, relative to the moving fortnightly average, were also defined.[58] Maximum hourly SO_2 levels at local monitoring stations (one in Harlem, two in Brooklyn) on each day were used as a measure of short-term SO_2 exposure. These peak levels correlated rather poorly with the daily mean levels used in previous analyses. Nevertheless, there was virtually no correlation between peak SO_2 level and asthma attendances. In Harlem during 1969, there were 59 days with peak SO_2 below 0.1 ppm (286 $\mu g/m^3$), of which 10 (17%) had unusually high numbers of attendances, and 34 (58%) had below average numbers of visits. The equivalent figures for 68 days with peak SO_2 above 0.3 ppm (858 $\mu g/m^3$) were 10 (15%) "high asthma days" and 36 (53%) "low asthma days". A similar pattern was evident on 17 days with peak levels above 0.5 ppm (1,430 $\mu g/m^3$). This carefully analysed study provides no support for an acute effect of short-term SO_2 exposure on asthma admissions, even at maximum hourly levels well in excess of current WHO guidelines (350 $\mu g/m^3$, 0.123 ppm).

Melbourne, Australia

8.39   The same method of defining days with an unusually high number of asthma attacks was used in a more recent study of 4,766 emergency attendances for asthma at the Royal Children's Hospital, Melbourne, during 1989.[59] Levels of fine particles were expressed in terms of a locally derived airborne particulate index (API), which correlated highly with measures of respirable particles (PM_{10} and $PM_{2.5}$). Although there was a significantly greater number of asthma attendances on days with a high API, within each season, there was virtually no relationship between high API days and unusually high asthma days. The latter represents a more rigorous but statistically more conservative analysis. The authors also examined the relationship between asthma attendances and high ozone levels (hourly maximum >180 $\mu g/m^3$, 90 ppb). No association was found for the number of asthma attendances, nor for unusually high asthma days.

Saint Nazaire, France

8.40   A smaller study of 372 hospital admissions for asthma among residents of the Loire estuary over an 18 month period found a significant correlation between daily asthma admissions and levels of Black Smoke, particularly among children.[60] There was a doubling of the admission rate among the under 15 age group on days when the smoke level was most elevated, but no significant correlation between asthma admissions and levels of SO_2. It is not clear how seasonal variations were controlled in this study, if at all. Caution is required in interpreting a finding from one subgroup in isolation.

Oulu, Finland

8.41   Another small study of 232 emergency visits to Oulu University Hospital for asthma during October 1985 to September 1986 found significant positive correlations between asthma incidence and levels of TSP, SO_2 and NO_2 on the same day.[61] Lagged pollutant concentrations were less highly correlated with asthma incidence. In a *weekly* analysis, NO_2 levels emerged as the only significant independent predictor of asthma visits, particularly in winter. It is impossible to properly evaluate the magnitude of these effects as no adjustment was made for seasonality, and both asthma attacks and high levels of each pollutant were more common in winter. Their statistical significance is likely to have been exaggerated because autocorrelation was not controlled and the authors inappropriately assumed Gaussian errors in the number of attendances which averaged less than one per day. The dose-response gradient derived for NO_2 is implausible at the low levels of exposure encountered (annual mean 13 $\mu g/m^3$, 7 ppb; range 0–69 $\mu g/m^3$, 0–37 ppb) and should be viewed with extreme caution in the light of these methodological weaknesses and other findings (see earlier discussion of the London NO_2 episode).

Helsinki, Finland

8.42   A second Finnish study, based on a much larger case-series, analysed admissions to Helsinki University Central Hospital during 1987–89.[62] Over this three-year period there were 4,209 asthma admissions (about 4 per day), of which one-third were children. After adjustment for temperature, daily admissions were significantly and positively correlated with levels of SO_2, NO and NO_2 (all present at low levels) and with TSP (annual mean 77 $\mu g/m^3$, with higher levels in the winter). Asthma visits were not associated with ozone levels (ranging up to 90 $\mu g/m^3$ in summer). As in the Oulu study, the statistical methodology is flawed in that no adjustment was made for short-wave autocorrelation or for seasonal cycles (other than those related to temperature).

Birmingham, UK

8.43 The only time series analysis hitherto published from the UK is a recent report based on admissions for asthma to Birmingham hospitals during April 1988 to March 1990.[48] Daily counts (average 5.4 per day) were related to 24-hour averages of Black Smoke and SO_2 levels at 7 sites in the city. The mean and maximum daily smoke levels were 13 $\mu g/m^3$ and 188 $\mu g/m^3$, respectively. The corresponding figures for SO_2 were 39 $\mu g/m^3$ and 126 $\mu g/m^3$ (14 and 44 ppb). Pollutant effects were adjusted for temperature, barometric pressure and humidity, and the possibility of short-wave autocorrelation was examined. Separate regression models were fitted for each pollutant in each of four seasons and lags of zero and two days were applied to the daily admission data. Weekly numbers of admissions were regressed against weekly means of pollutant levels and meteorological variables. This approach generated a large number of regression models, among which a significant correlation emerged only in winter for daily smoke (with 2-day lag) and weekly SO_2. The authors estimated that a rise in winter smoke level of 20 $\mu g/m^3$ would result in one additional asthma admission per day (a relative increase of 19%, 95% CI: 2% to 34%). An important omission in this study was failure to adjust for the effects of the November 1989 influenza epidemic, which increased asthma admissions during early winter, a time of increased smoke and SO_2 pollution. It is, therefore, almost certain that the magnitude of the pollutant effects were overestimated. The positive associations with asthma admissions were only of marginal statistical significance, as might be expected by chance alone when, as in this anaylsis, multiple comparisons have been made. This single UK study, therefore, provides no strong evidence of an effect of particulate or SO_2 pollution on the timing of asthma attacks.

Seattle, USA

8.44 A recent report focusing specifically on the relationship of asthma incidence to particles of respirable size (less than 10 μm in diameter) was based on 2,809 emergency room visits to eight hospitals in the Seattle area during the period September 1989 to September 1990.[49] This analysis included state-of-the-art statistical methods, using autoregressive models to control for short-wave autocorrelation; Poisson errors appropriate to small numerators, with allowance for additional (extra-Poisson) errors; adjustment for seasonal and other variations in asthma occurrence by fitting long-wave cycles, day-of-the-week effects, temperature (both as a continuous variable and additionally as "cold days"); and sensitivity analysis to confirm that changes to the model structure did not materially alter the conclusions. Levels of PM_{10}, SO_2 and ozone (during 5 summer months) were each obtainable only from one site which was different for each pollutant. The seasonal pattern of particulate pollution (high in winter, low in summer) differed from the seasonality of asthma (peaks in March and September). However, after allowance for seasonality, temperature and day of the week, a statistically significant positive correlation was observed between asthma visits and PM_{10} levels, the relationship being strongest for the average PM_{10} concentration over the previous 4 days. The dose-response relationship was graded with no evidence of a threshold within the range of PM_{10} levels observed (10–60 $\mu g/m^3$). It was of similar strength in each season and among children and young adults, but no association was found in the 65+ age group. Among the under-65s, a 30 $\mu g/m^3$ increase in PM_{10} (4-day average) was associated with a 12% (95% CI 4% to 20%) relative increase in asthma attendances. There was virtually no association between SO_2 levels and asthma incidence. Ozone levels were negatively associated with asthma visits, but this relationship was not statistically significant (for details see Table 8.1). This study was carefully and thoroughly analysed but was based on a limited period of observation and single monitoring sites for each pollutant. It is possible that during the single year of observation PM_{10} levels were coincidentally high at the time of viral epidemics, or that some local source of particulate pollution (such as wood smoke) was of importance as a source of indoor pollution. The consistency of the PM_{10} effect in each season argues against, but does not entirely exclude, such confounding effects.

Utah Valley, USA

8.45 An unusual opportunity to study the effects of changing levels of particulate pollution arose in Utah Valley during the late 1980s. The main source of pollution in this community was a steel mill which accounted for over half of the particulate emissions and more than 80% of emissions of sulphur oxides, nitrogen oxides and hydrocarbons.[63] The mill was closed for 13 months (August 1986 to September 1987) due to an industrial dispute. Levels of PM_{10} were monitored in the area continuously from 1985 onwards, and were characterised by high levels in December and January in association with temperature inversions and still weather conditions. The mean PM_{10} level during the winter of the mill closure was 51 $\mu g/m^3$, compared to 90 $\mu g/m^3$ and 84 $\mu g/m^3$ in the preceding and subsequent winters. The US EPA guideline level of 150 $\mu g/m^3$ for PM_{10} was exceeded frequently in the winters when the mill was open, but not at all when it was closed. Concentrations of ambient SO_2, NO_x and ozone (in the winter months) were said to be low, although monitoring of these pollutants had been discontinued at the time of the mill closure. The community was also unusual in that most of the residents were of the Mormon faith, which forbids smoking, so only 6% of the population were current smokers. Outdoor air pollution would, therefore, have contributed a greater part of personal exposure to particles than in other areas.

8.46 The number of hospital admissions for asthma and bronchitis in this community was relatively small: about one per day.[63] Two conventional time series analyses of respiratory admissions have been published.[50,64] Each of these used the *monthly* number of admissions as the outcome variable and found a significant positive correlation between the mean PM_{10} level in the *previous* month (lagged PM_{10}) and admissions for bronchitis and asthma (combined) in both children and adults. These analyses adjusted for mean monthly temperature but have been criticised for failing to control for other seasonal factors, in particular the winter epidemic of respiratory syncytial virus infection which coincided with the peak PM_{10} levels.[47] Furthermore, a reanalysis of the data for all respiratory admissions combined suggests that the effects quoted are attributable to three "outliers" (months with high lagged PM_{10} levels and high rates of admission), whereas within the bulk of the data little correlation is apparent between respiratory admission rates and PM_{10} (either lagged or unlagged).[47] These time series analyses did not analyse asthma separately and because of the limitations mentioned above they contribute little to the quantification of air pollution effects.

8.47 A more scientifically interesting and statistically rigorous approach has been to compare the illness rates during the autumn and winter seasons before, during and after the period of the mill closure. The first such report was based on comparisons entirely within Utah Valley.[64] Admissions for bronchitis and asthma (combined) during the winter of the closure were markedly reduced among children (by about two-thirds), but not among adults. Autumn admissions were also reduced among children (by about 50%) but not among adults. The changes among children were statistically highly significant, but could be spurious if, by coincidence, epidemics of viral infection had been mild in the study area during the year of the closure. Economic hardship resulting from the closure was considered an unlikely explanation as other illness categories were unaffected.

8.48 In a further analysis,[50] admission rates in two other areas (Salt Lake Valley and Cache Valley) were compared to those for Utah Valley for the 4-year period April 1985 to March 1989. Regional epidemics of infectious disease could have been expected to affect each community equally. Cache Valley was chosen as a relatively unpolluted control area but is of limited scientific value as pollutant data were sparse and the number of asthma admissions was low (35 per year). In Salt Lake Valley, winter PM_{10} levels were similar before, during and after the Utah Valley steel mill closure, and the catchment population was large, generating 294 asthma admissions per year, compared to 150 per year in Utah Valley. Annual admission rates for asthma (all ages) during the period of the mill closure were reduced by 4% in Salt Lake Valley and by 12% in Utah Valley. Neither of these effects was statistically significant, so the difference between them cannot be. However, among children aged 0–5 years in Utah Valley, there was a highly significant reduction in admission rates for asthma and bronchitis (each reduced by about 50%) during the year of the mill closure. The corresponding changes in Salt Lake Valley children were small and non-significant: reductions of 15% for asthma and 11% for bronchitis. Unfortunately, the statistical significance of the difference between the two areas was not assessed, although it seems likely from the data presented that it would be statistically significant for pre-school children.

Hong Kong

8.49 The association of asthma admissions with levels of respirable particles (PM_{10}), total suspended particulates (TSP) and gaseous pollutants (SO_2, NO_x and ozone) was studied in Hong Kong during 1983–89.[65] Over this period, asthma admission rates rose in children aged 1–4 year, fell among 5 to 14 year-old children and changed little in adults. Unfortunately, for administrative reasons, the admissions were collated on a quarterly basis, so short-term variations in asthma incidence could not be analysed. At this level of aggregation, it is difficult to distinguish seasonal and pollution effects with confidence. However, after adjusting for season and year, the authors report significant positive correlations between TSP and asthma admission rates at all ages over 1 year, and significant *negative* correlations of similar strength between SO_2 and asthma admission rates in all ages. Concentrations of respirable particles were most strongly correlated with asthma in the 1–4 age-group. There were also *negative* correlations in NO_x in all age groups from 1 year upwards, but weak correlations with NO_2 and ozone levels. The causal interpretation of these findings is uncertain, but it is of interest that the correlations of levels of each pollutant with admission rates were consistent across the age range, with the exception of PM_{10}, which was more strongly correlated in younger children.

"Summer haze" pollution

California, USA: students

8.50 An early study of acute health effects from photochemical smog was based on Californian universities (5 in Los Angeles, 2 in San Francisco), each closely adjacent to an air pollution monitoring site.[66] Students attending health centres during the period October 1970 to May 1971 within 10 days of the onset of specified respiratory illnesses were included in the analysis. Among 11,659 episodes of respiratory illness, asthma and hay fever were among the three diagnoses whose timing was *least* highly correlated with any of the pollutants measured (particles, SO_2, NO_2, NO, total NO_x, CO, hydrocarbons, total oxidants). No more specific analyses of asthma attack rates are presented.

California, USA: children

8.51 In a later study of 2,139 emergency room visits for asthma and bronchiolitis at one children's hospital in Los Angeles,[67] daily attendances during August 1979 to January 1980 (approximately 12 per day) were analysed in relation to pollutants measured at five neighbouring monitoring sites and aeroallergen levels measured at two sites. Correlations with each pollutant separately were significantly positive for aeroallergens, coefficient of haze, hydrocarbons and nitrogen oxides, and significantly negative for SO_2 and ozone. There were non-significant correlations with total particles (positive) and sulphates (negative). Temperature and relative humidity were not controlled in these analyses, but each were negatively correlated with asthma attendances. A factor analysis of the pollution and climatic variables suggested three pollution patterns with independent effects on asthma attendances. The first, characterised by dry, hazy conditions, "Santa Ana wind" and raised levels of hydrocarbons and NO_x, was associated with an increased incidence of asthma attacks. The second, characterised by high temperature, ozone and SO_2 levels, was negatively correlated with the attack rate. The third, related to high aeroallergen levels, was positively correlated with asthma incidence. The relative influence of each factor was much the same, but unfortunately their statistical significance is not reported, nor can a quantitative estimate of the pollution effects be easily derived. No attempt was made to control for the underlying seasonality of asthma, but the authors comment that the autumn peak in admissions in that area coincided with the seasonal increase in particulate pollution, haze and nitrogen oxides, and a decrease in ozone levels. They point out that the increased incidence of respiratory infections at this time could have generated spurious correlations with pollution and meteorological variables.

Georgia, USA

8.52   During the three-month period June to August 1990, 609 emergency clinic visits by children in Atlanta, Georgia, were related to levels of ozone measured on the previous day at two local monitoring sites.[68] Maximum hourly ozone levels exceeded 110 ppb (220 μg/m³) on six days, and the mean daily number of visits on the following six days was 8.5, compared to 6.4 per day otherwise. Ozone levels were correlated with pollen counts (r=0.26) and PM_{10} levels (r=0.47). After adjustment for these factors plus temperature and day of the week by autoregressive Poisson modelling, the incident effect of "high" (>110 ppb, 220 μg/m³) ozone levels was estimated as a 33% increase in asthma visits on the following day (95% CI: –9% to +92%). This finding is not statistically significant, although from the data presented there is evidence of a dose-response trend across the range of ozone levels, which is not tested for significance. The independent effect of PM_{10} levels was not statistically significant (2% increase per 10 μg/m³ increase in PM_{10}, 95% CI: –4% to +13%).

Ontario, Canada

8.53   A large and influential set of analyses of hospital admissions for asthma in southern Ontario were published by Bates and Sizto during the 1980s.[69–71] Information was obtained on admissions to 79 hospitals serving a population of 5.9 million people over the years 1974 and 1976–83. Forty three per cent of the asthma admissions were among children under 15 years of age. Analysis was restricted to the winter months of January and February and the summer months of July and August. The influence of seasonal fluctuations in asthma incidence was further controlled by using as the outcome variable the deviation of each day's admissions from the average daily admission in the same week (within the same season and same year). These daily deviations (based on about 20 admissions per day) were correlated with the levels of ozone, SO_2, NO_2 and coefficient of haze measured hourly at 17 monitoring stations in the region. The average of the hourly maxima across all sites on any given day was used. In addition, airborne sulphates were measured every sixth day and a 24-hour average was used. Temperature, humidity and pollution data were analysed with lags of 0, 1 and 2 days.[70]

8.54   During the winter months (11,186 admissions), only temperature was significantly (positively) correlated with asthma admissions in all ages. There was a significant *inverse* correlation of NO_2 levels with winter asthma admissions in children. During the summer months, no correlation was found with temperature or humidity, but there were significant positive correlations of asthma admissions with levels of SO_2 (lagged 2 days); ozone (lagged 1 day); and sulphates (lagged 1 day). No significant correlations emerged when the analysis was restricted to children. There was a high degree of cross-correlation between summer levels of ozone, sulphates, SO_2, NO_2 and temperature on the same day, so the independent effects of each pollutant could not be estimated with any certainty. However, on days with the highest hourly ozone level in each season ("high ozone" days), there were approximately 7% more admissions for respiratory illnesses than on "low ozone" days (days with the lowest hourly ozone level in each season). Unfortunately, no corresponding estimate is presented specifically for asthma admissions. However, a later paper, based on the same geographical area estimated that a 50 ppb (100 μg/m³) increase in maximum hourly ozone during July and August was associated with a significant 8.3% increase in admissions for childhood asthma on the subsequent day.[72]

8.55   In an earlier paper,[71] the authors of the Ontario study discuss which pollutant might have been responsible for the increase in respiratory admissions on high ozone days. They point out that no excess of admissions for respiratory illnesses occurred during June 1983. This was a month excluded from the previous analysis, but characterised by unusually high ozone levels. More specifically, there were 913 admissions for asthma in June 1983 compared with an average number of 992 in that month in other years. Sulphates and SO_2 were considered an unlikely cause as no correlation was found with these pollutants in winter, despite similar concentrations of sulphates and higher levels of SO_2 than in summer. They speculated that aerosolised sulphuric acid might be responsible for the "acid summer haze" effect, but were unable to investigate this further due to paucity of pollutant data for airborne acid.

8.56 Direct measurements of acid aerosol were included in a further analysis of asthma admissions to 22 acute care hospitals in Toronto during the summers of 1986, 1987 and 1988.[73] There were approximately 9 asthma admissions per day. Time-series methods were employed to remove seasonal cycles and day-of-week effects. After adjustment for temperature, there were significant positive correlations of daily asthma admissions with ozone (same day) and acid (same day). Daily levels of these two pollutants were correlated (r=0.51), but less strongly than ozone and particles (r=0.75) or ozone and sulphates (0.82). In two-pollutant models, ozone was a more significant predictor of asthma admissions than airborne acid, sulphates or particulate pollution, each of which were non-significant (at the 10% level) after adjustment for ozone. The independent effect of ozone was estimated as a 1.5% increase (95% CI: 0.1% to 3.0%) in admission rate per 10 $\mu g/m^3$ (5 ppb) increase in hourly ozone concentration on the same day. This corresponds to a 32% increase in asthma admission rate on days of maximum ozone level (159 ppb, 318 $\mu g/m^3$), compared to days of average ozone concentration (70 ppb, 140 $\mu g/m^3$). This thorough analysis offers no support for the hypothesis that airborne acid, rather than ozone itself, is responsible for the association of asthma admissions with levels of photochemical air pollution in Ontario. It is consistent with the modest effect of ozone suggested by other Ontario studies.

Vancouver, Canada

8.57 In order to further disentangle the health effects of ozone and airborne acid or sulphates, Bates and colleagues carried out a similar study in Vancouver, where summer ozone peaks were not associated with acid or sulphate pollution.[74] In this population of about 1 million, emergency room attendances rather than hospital admissions were used in order to increase the number of daily asthma attacks included in the analysis. Over the period July 1984 to October 1986, there were about ten attendances with asthma per day, 40% of them among children. The seasonal pattern was similar to that in Britain, with a late September peak and no winter excess of asthma attacks. Ozone levels peaked in July and August and had fallen substantially before the September peak in asthma incidence. There was an autumnal rise in SO_2, NO_2 and coefficient of haze, but this was preceded by the rise in asthma attendances. When the seasonal fluctuations were removed using the same method as in Ontario,[65] there was no significant correlation of asthma attendances with ozone levels in any age group. Airborne sulphates were significantly and positively correlated with asthma in each age group (though with different lags) during the summer, and with asthma among over-60s (only) in winter. A broadly similar pattern emerged for sulphur dioxide. As in the Ontario study, no quantitative estimates of these effects are presented.

New Jersey, USA

8.58 A study designed to replicate the findings of Bates and colleagues was carried out in New Jersey, a region with higher summertime ozone levels than southern Ontario or Vancouver.[51] Emergency room visits for asthma at nine hospitals during two summers (May to August 1988 and 1989) were analysed in relation to average levels of ozone measured between 10 am and 3 pm each day at five monitoring stations in the region. The analysis adjusted for autocorrelation by restriction to the summer season, adjusting for temperature effects and using an autoregressive model. Only after adjusting for temperature (which was negatively correlated with asthma attendances, but positively correlated with ozone levels) did ozone (lagged by 1 day) emerge as a significant predictor of asthma attacks. The independent effect of ozone was estimated as a 2.4% (95% CI: 0.7%–4.0%) proportionate increase in asthma attendances per 10 ppb (20 $\mu g/m^3$) increase in ozone concentrations on the previous day. These results should be interpreted with some caution because it is not clear that the adjustment for temperature fully removed the influence of month-to-month variations in asthma incidence.

New York and Buffalo, USA

8.59   A second study with similar aims was carried out over the same two summers (1988 and 1989) in New York city, Buffalo and surrounding areas.[52] This large population of 11 million people generated about 65 emergency admissions for asthma each day to hospitals in New York state. The seasonal variation in asthma incidence was controlled by restricting the analysis to the summer months (June to August) and fitting long-wave cycles to the data. Adjustments were also made for day of the week effects and short-wave autocorrelation of the residual variations in asthma incidence were assessed and found to be minimal. Measures of hourly ozone, daily sulphates and daily acid aerosol were available from three monitoring sites. The maximum daily ozone level was used in the analysis. Daily levels of the three pollutants were highly cross-correlated and the independent effects could not be distinguished in the analysis. Each was positively correlated with asthma attendances on the same or subsequent days. Quantitative results are presented for New York city (49 admissions per day) for 1988 only. These are said to be broadly consistent with the results for 1989 and for other areas studied. After adjustment for temperature, the independent effect of ozone (lagged 1 day) was estimated as a 1.7% (95% CI: 0.3%–3.0%) proportionate increase in asthma attendances per 10 ppb (20 $\mu g/m^3$) increase in hourly maximum ozone on the previous day. This corresponded to a 23% relative increase in asthma admissions on a day of maximal ozone levels (412 $\mu g/m^3$, 206 ppb) compared to a day of average ozone levels (138 $\mu g/m^3$, 69 ppb). An effect of similar relative magnitude was obtained for the corresponding analysis of airborne acid (32% increase) and sulphates (19% increase). Each of these pollutants was found to fit best with zero lag period. This is a carefully analysed study which generated associations with ozone of similar magnitude to those in New Jersey and Toronto over the same period.[51] Unfortunately, the addition of acid aerosol measurements did not clarify which specific pollutant was responsible for the "acid summer haze" effects.

Comment

8.60   Many of the time series analyses which have been reported are based on relatively short periods of observation (one or two years). Although long-wave cycling in asthma incidence has been controlled with a varying degree of rigour in different studies, it is difficult on the basis of a single year's data to exclude a coincidental rise in pollution levels at the time when asthma attacks might be on the increase for other reasons (such as epidemics of respiratory virus infection in autumn or winter, or aeroallergen exposure in spring, summer and autumn).

8.61   Only a minority of the studies reviewed published evidence in a form which permitted an estimation of the dose-response relationship to individual pollutants. It should be noted that quantitative estimates were rarely quoted when no significant correlation was observed between asthma incidence and pollutant levels. Thus, the results preferentially include "positive" associations and may overestimate the dose-response gradient.

8.62   Asthmatic patients are known to be unusually sensitive to sulphur dioxide, yet this is one pollutant when the time series analyses consistently show no association with the timing of asthma attacks, even when peak (hourly maximum) exposures were considered. The evidence is much less consistent with regard to particles. This may reflect different particle composition in different areas, or a correlation of outdoor particles with indoor pollutant exposure which varies according to the fuel used for domestic heating. On the other hand, the results from Seattle (affected by wood smoke) and Utah Valley (affected by industrial emissions) are fairly consistent, supporting the recent suggestion that it is the density of respirable particles, rather than their chemical composition, which is the influential factor in determining health effects.[75]

8.63   The comparative analyses from Utah and Salt Lake Valleys of admission rates in areas affected and unaffected by a transient reduction in winter particulate pollution[50] are consistent with a causal association between exposure to high levels of PM_{10} pollution and the occurrence of wheezing illnesses in young children. However, the results for asthma across the whole age range are less striking (see Table 8.1), and caution should be adopted in interpreting the findings for one subgroup presented in isolation. It is of interest that in Hong Kong the strongest correlation between quarterly asthma admission rates and levels of respirable particles was also found in the 1–4 age-group,[65] and in Saint Nazaire the correlation was stronger in children than in adults.[60] In contrast, there was little age differential in the dose-response gradient relating PM_{10} to emergency asthma visits by children and adults in Seattle.[49]

8.64   There is fairly consistent evidence that the type of "summer haze" affecting southern Canada and the eastern USA is associated with a modest short-term increase in the incidence of asthma attacks. The causal agent, however, remains in doubt. On the western coast of North America, ozone levels were either uncorrelated with asthma incidence (as in Vancouver and Seattle) or inversely correlated with daily asthma attack rates (Los Angeles). Acid aerosol was considered as a possible agent in southern Ontario, but direct measurements of airborne acid in New York[52] were no more closely correlated with asthma admissions than were ozone or sulphate levels. Airborne sulphates form a major part of the fine particle fraction in northeast USA and it has been suggested that the effects of summer haze may be mainly attributable to these respirable particles.[75] However, the only summer haze study to measure fine particles directly[51] found no relationship between PM_{10} levels and the timing of asthma attacks.

8.65   The only British study[48] relates to smoke and SO_2 pollution during a period which included a major influenza epidemic. After excluding the results for winter months, which may be confounded by this epidemic, there was little effect of pollution levels on daily asthma admissions (see Table 8.1). This contrasts with the results obtained for particulate pollution in the USA, and suggests that the American findings should be extrapolated with some caution to the British situation. Even greater problems are apparent in generalising the results of summer haze studies, since here there is inconsistency within the American continent.

Conclusions

8.66   Asthma is a disease characterised by considerable short-term variability due to factors other than air pollution: epidemics of virus infection; aeroallergen exposure; and perhaps abrupt climatic changes (see Chapter 7). By comparison, the effects of short-term fluctuations in measured air pollutants are subtle and difficult to quantify with certainty.

8.67   There is no direct evidence from population based studies of an association between the timing of asthma attacks and day-to-day fluctuations in the type of air pollution currently encountered in the UK. This conclusion reflects the paucity of time series data from the UK and the inconsistency of results obtained elsewhere. There is a clear need for substantial and statistically rigorous analyses of the relationship between asthma incidence and daily pollutant levels in Britain.

8.68   None of the marked variations in the occurrence of asthma attacks which have been documented were related to changes in the level of measured pollutants. These well defined "asthma epidemic days" are usually related to high concentrations of aeroallergens.

8.69   In most of the severe episodes of "old-fashioned" smoke and SO_2 pollution, there was little evidence of clinical complications among young asthmatic patients, although older patients with all forms of chronic obstructive airways disease were adversely affected. Asthma does not appear to be the most sensitive indicator of adverse health effects from the type of winter smog encountered in Britain in the past. However, there is only one study which related to a well defined episode of "modern" vehicle-related pollution, in London, December 1991. Analyses to date do not support the widespread concern regarding an epidemic of wheezing illness in the capital at this time, and, as in previous smog episodes, other respiratory outcomes, particularly among the elderly, seem to have been affected to a greater degree.

References

1. Carroll RE. Environmental epidemiology. V. Epidemiology of New Orleans epidemic of asthma. Am J Public Health 1968; 58:1677–1683.

2. Kenline PA. October 1963 New Orleans asthma study. Arch Environ Health 1966; 12:295–304.

3. Salvaggio J, Kawai T, Seabury J. New Orleans epidemic asthma: semiquantitative aerometric sampling, epidemiologic and immunologic studies. Chest 1973; 63(Suppl) 14S–15S.

4. Antó JM, Sunyer J. Asthma Collaborative Group of Barcelona. A point source asthma outbreak. Lancet 1986; i:900–903.

5. Antó JM, Sunyer J, Rodriquez-Roisin R, Suarez-Cervera M, Vacquez L. Community outbreaks of asthma associated with inhalation of soybean dust. N Engl J Med 1989; 320:1097–1102.

6. Sunyer J, Antó JM, Rodrigo MJ, Morell F. Case-control study of serum immunoglobulin-E antibodies reactive with soybean in epidemic asthma. Lancet 1989; i:179–182.

7. Ussetti P, Roca J, Augusti AGN, Montserrat JM, Rodriguez-Roisin R, Agusti-Vidal A. Another asthma outbreak in Barcelona: role of oxides of nitrogen. Lancet 1984; i:156.

8. Antó JM, Sunyer J, Plasencia A. Nitrogen dioxide and asthma outbreaks [Letter]. Lancet 1986; ii:1096–1097.

9. Castellsague J, Sunyer J, Salz M, Murillo C, Antó JM. Effect of air pollution in asthma epidemics caused by soya bean dust [Abstract]. Eur Respir J 1992; Suppl 15:413S.

10. Packe GE, Ayres JG. Asthma outbreak during a thunderstorm. Lancet 1985; 2:199–204.

11. Harries MG, Lacey J, Tee RD, Cayley GR, Newman Taylor AJ. *Didymella exitialis* and late summer asthma. Lancet 1985; i:1063–1066.

12. Egan P. Weather or not. Med J Aust 1985; 142:330.

13. Bellomo R, Gigliotti P, Treloar A, Holmes P, Suphioglu C, Singh MB, Knox B. Two consecutive thunderstorm associated epidemics of asthma in the city of Melbourne. The possible role of rye grass pollen. Med J Aust 1992; 156:834–837.

14. Suphioglu C, Singh MB, Taylor P, Bellomo R, Holmes P, Puy R, Knox RB. Mechanism of grass-pollen-induced asthma. Lancet 1992; 339:569–572.

15. Murray V, Venables K, Laing-Morton T, Partridge M, Thurston J, Williams D. Epidemic of asthma possibly related to thunderstorms. BMJ 1994; 309:131–132.

16. Higham JH. Asthma trends: thunderstorm peak in Luton [Letter]. BMJ 1994; 309:604.

17. Figley KD, Elrod RH. Endemic asthma due to castor bean dust. JAMA 1928; 90:79–82.

18. Mendes E, Cintra AU. Collective asthma, simulating an epidemic, provoked by castor bean dust. J Allergy 1954; 25:253–259.

19. Ordman D. Outbreak of bronchial asthma in South Africa affecting more than 200 persons caused by castor bean dust from an oil processing factory. Int Arch Allergy 1955; 7:10–24.

20. Ministry of Health. Mortality and morbidity during the London fog of December 1952. London: HMSO, 1954.

21. Martin E. Mortality and morbidity statistics and air pollution. Proc R Soc Med 1964; 57:969–975.

22. Fry J. Effects of a severe fog on a general practice. Lancet 1953; i:235.

23. Fry J, Dillane KB, Fry L. Smog: 1962 v 1952 [Letter]. Lancet 1962; ii:1326.

24. Firket J. Sur les causes des accidents survenues dans la valle de la Meuse, lors des brouillards de Decembre 1930. Bull Acad R Med Belg 1931; 11:683.

25. Schrenk HH, Heinmann H, Clayton GD, Gafafar WM, Wexler H. Air pollution in Donora PA: epidemiology of the unusual smog episode of October 1948. Public Health Bulletin No 306. Washington DC: Public Health Service, 1949.

26. Greenburg L, Field F, Reed JI, Erhardt CL. Air pollution and morbidity in New York City. JAMA 1962; 182:161–164.

27. Greenburg L, Erhardt CL, Field F, Reed JI. Air pollution incidents and morbidity studies, New York city. Arch Environ Health 1965; 10:351–356.

28. Glasser M, Greenburg L, Field F. Mortality and morbidity during a period of high levels of air pollution. New York, Nov 23 to 25, 1966. Arch Environ Health 1967; 15:684–694.

29. Chiaramonte LT, Bongiorno JR, Brown R, Laano ME. Air pollution and obstructive respiratory disease in children. New York State J Med 1970; Feb:394–398.

30. Becker WH, Schilling FJ, Verma MP. The effect on health of the 1966 seaboard air pollution episode. Arch Environ Health 1968; 16:414–419.

31. Baxter PJ, Ing R, Falk H, Plikaytis B. Mount St Helens eruptions: the acute respiratory effects of volcanic ash in a North American community. Arch Environ Health 1983; 38:138–143.

32. Wichmann HE, Mueller W, Allhoff P, Beckmann M, Bocter N, Csiasky MJ, Jung M, Molik B, Schoeneberg G. Health effects during a smog episode in West Germany in 1985. Environ Health Perspect 1989; 79:89–99.

33. Ayres J, Fleming D, Williams M, McInnes G. Measurement of respiratory morbidity in general practice in the United Kingdom during the acid transport event of January 1985. Environ Health Perspect 1989; 79:83–88.

34. Anderson HR, Limb ES, Bland JM, Ponce de Leon A, Strachan DP, Bower J. The health effects of an episode of nitrogen dioxide air pollution in London December 1991. Thorax 1995 [In press].

35. Department of Health. Advisory Group on the Medical Aspects of Air Pollution Episodes. Third Report: Oxides of Nitrogen. London: HMSO, 1993.

36. Ransom MR, Pope CA. Elementary school absences and PM_{10} pollution. Environ Res 1992; 58:204–219.

37. Pönkä A. Absenteeism and respiratory disease among children and adults in Helsinki in relation to low-level air pollution and temperature. Environ Res 1990; 52:34–46.

38. Romieu I, Lugo MC, Velasco SR, Sanchez S, Meneses F, Hernandez M. Air pollution and school absenteeism among children in Mexico City. Am J Epidemiol 1992; 136:1524–1531.

39. Levy D, Gent M, Newhouse MT. Relationship between acute respiratory illness and air pollution levels in an industrial city. Am Rev Respir Dis 1977; 116:167–173.

40. Goldsmith JR, Griffith HL, Detels R, Beeser S, Neumann L. Emergency room admissions, meteorologic variables and air pollutants: a path analysis. Am J Epidemiol 1983; 118:759–778.

41. Samet JM, Bishop Y, Speizer FE, Spengler JD, Ferris BG. The relationship between air pollution and emergency room visits in an industrial community. JAPCA 1981; 31:236–240.

42. Diaz-Caneja N, Gutierrez I, Martinez A, Mattoras P, Villar E. Multivariate analysis of the relationship between meteorological and pollutant variables and the number of hospital admissions due to cardio-respiratory diseases. Environ Int 1991; 17:397–403.

43. Mazumdar S, Sussman N. Relationship of air pollution to health: results from the Pittsburgh study. Arch Environ Health 1983; 38:17–24.

44. Kardaun JWPF, van der Maas PJ, Habbema JDF, Leentvaar-Kuijpers A, Rijcken B. Incidence of diseases of the lower respiratory tract in family practice and low level air pollution. Family Practice 1989; 6:86–91.

45. Schwartz J, Spix C, Wichmann HE, Malin E. Air pollution and acute respiratory illness in five German communities. Environ Res 1991; 56:1–14.

46. Sunyer J, Antó JM, Murillo C, Saez M. Effects of urban air pollution on emergency room admissions for chronic obstructive pulmonary disease. Am J Epidemiol 1991; 134:277–286.

47. Lipfert FW. A critical review of studies of the association between demands for hospital services and air pollution [Review]. Environ Health Perspect 1993; 101(Suppl 2):229–268.

48. Walters S, Griffiths RK, Ayres J. Temporal association between hospital admissions for asthma in Birmingham and ambient levels of sulphur dioxide and smoke. Thorax 1994; 49:79–83.

49. Schwartz J, Slater D, Larson TV, Pierson WE, Koenig JQ. Particulate air pollution and hospital emergency room visits for asthma in Seattle. Am Rev Respir Dis 1993; 147:826–831.

50. Pope CA. Respiratory hospital admissions associated with PM_{10} pollution in Utah, Salt Lake and Cache Valleys. Arch Environ Health 1991; 46:90–97.

51. Cody RP, Weisel CP, Birnbaum G, Lioy PJ. The effect of ozone associated with summertime photochemical smog on the frequency of asthma visits to hospital emergency departments. Environ Res 1992; 58:184–194.

52. Thurston GD, Ito K, Kinney PL, Lippmann M. A multi-year study of air pollution and respiratory hospital admissions in three New York State metropolitan areas: results for 1988 and 1989 summers. J Expo Anal Environ Epidemiol 1992; 2:429–450.

53. Derrick EH. A comparison between the density of smoke in the Brisbane air and the prevalence of asthma. Med J Aust 1970; 2:670–675.

54. Girsch LS, Shubin E, Dick C, Schlaner FA. A study on the epidemiology of asthma in children in Philadelphia: the relation of weather and air pollution to peak incidence of asthmatic attacks. J Allergy 1967; 39:347–357.

55. Goldstein IF, Duberg E. Air pollution and asthma: search for a relationship. JAPCA 1981; 31:370.

56. Ribon A, Glasser M, Sudhivoraseth N. Bronchial asthma in children and its occurrence in relation to weather and air pollution. Ann Allergy 1972; 30:276–281.

57. Goldstein IF, Weinstein AL. Air pollution and asthma: effects of exposures to short term sulfur dioxide peaks. Environ Res 1986; 40:332–345.

58. Goldstein IF, Rausch LE. Time series analysis of morbidity data for assessment of acute environmental health effects. Environ Res 1978; 17:266–275.

59. Rennick GJ, Jarman FC. Are children with asthma affected by smog? Med J Aust 1992; 156:837–841.

60. Chailleux E, Guyon C, Taddei F, Bouillard J, Pioche D. Asthma and air pollution. A study of admissions to the hospital of Saint-Nazaire. Rev Mal Respir 1990; 7:563–568.

61. Rossi OVJ, Kinnula VL, Tienari J, Huhti E. Association of severe asthma attacks with weather, pollen and air pollutants. Thorax 1993; 48:244–248.

62. Pönkä A. Asthma and low level air pollution in Helsinki. Arch Environ Health 1991; 46:262–270.

63. Pope CA. Respiratory disease associated with community air pollution and a steel mill, Utah Valley. Am J Public Health 1989; 79:623–628.

64. Pope CA, Dockery DW, Spengler JD, Raizenne ME. Respiratory health and PM_{10} pollution. A daily time series analysis. Am Rev Respir Dis 1991; 144:668–674.

65. Tseng RY, Li CK, Spinks JA. Particulate air pollution and hospitalization for asthma. Ann Allergy 1992; 68:425–432.

66. Durham WH. Air pollution and student health. Arch Environ Health 1974; 28:241–254.

67. Richards W, Azen SP, Weiss J, Stocking S, Church J. Los Angeles air pollution and asthma in children. Ann Allergy 1981; 47:348–354.

68. White MC, Etzel RA, Wilcox WD, Lloyd C. Exacerbations of childhood asthma and ozone pollution in Atlanta. Environ Res 1994; 65:56–68.

69. Bates DV, Sizto R. Relationship between air pollutant levels and hospital admissions in Southern Ontario. Can J Public Health 1983; 74:117–122.

70. Bates DV, Sizto R. Air pollution and hospital admissions in Southern Ontario: the acid summer haze effect. Environ Res 1987; 43:317–331.

71. Bates DV, Sizto R. The Ontario air pollution study: identification of the causal agent. Environ Health Perspect 1989; 79:69–72.

72. Burnett RT, Dales RE, Raizenne ME, Krewski D, Summers PW, Roberts GR, Raad-Young M, Dann T, Brook J. Effects of low ambient levels of ozone and sulfates on the frequency of respiratory admissions to Ontario hospitals. Environ Res 1994; 65:172–194.

73. Thurston GD, Ito K, Hayes CG, Bates DV, Lippmann M. Respiratory hospital admissions and summertime haze air pollution in Toronto, Ontario: consideration of the role of acid aerosols. Environ Res 1994; 65:271–290.

74. Bates DV, Baker-Anderson M, Sizto R. Asthma attack periodicity: a study of hospital emergency visits in Vancouver. Environ Res 1990; 51:51–70.

75. Dockery DW, Pope CA. Acute respiratory effects of particulate pollution. Annu Rev Public Health 1994; 15:107–132.

Chapter 9

Air Pollution and the Prevalence of Asthma: Population Studies

Introduction

9.1   This chapter reviews the evidence relating measures of the prevalence of asthma or wheezing illness to measured levels of ambient air pollution, point sources of pollution, and proximity to motor traffic. Recent research from Germany is reviewed in some detail because it includes objective measures of asthma and allergy in areas with widely differing patterns of past air pollution exposure.

9.2   Studies which have related *changes* in asthma prevalence or severity to *changes* in air pollution exposure over a number of years are also reviewed here. Evidence from population studies relating measures of asthma occurrence to short-term fluctuations in air pollution are included in Chapter 8.

9.3   Earlier in this report, two questions were distinguished:

(a)   Does air pollution, of the type and level likely to be experienced in the UK, precipitate attacks of asthma among persons who already have the disease?

(b)   Does such air pollution exposure influence the proportion of the population who suffer from asthma or other allergic disease?

The evidence to be considered in this chapter is particularly relevant to the second of these issues.

Comparisons of high and low pollution areas

Methodological considerations

9.4   A number of studies have investigated the prevalence of respiratory symptoms in areas of differing outdoor pollution exposure. Unfortunately, relatively few of these contain information specific to asthma or other wheezing illnesses. The majority of such studies are based on a comparison between a few areas (usually two) with different pollution profiles. There are important limitations inherent in geographical comparisons based on a small number of areas:

(a)   Differences in prevalence may arise from variation in factors other than air pollution, including artefacts due to interpretation of questions relating to symptoms or diagnosis.

(b)   The specific pollutant or pollutant mix which may be responsible for any difference cannot be defined.

(c)   The shape of any dose-response relationship between pollutant levels and health effects cannot be estimated.

9.5   A further theoretical difficulty is that the statistical significance of the prevalence difference cannot be assessed rigorously without supplementary information on the underlying geographical variability of asthma. However, the evidence from the UK (reviewed in Chapter 7) suggests that there is little spatial variability in asthma prevalence between small geographical areas such as towns or cities, after allowance for broad regional variations, so this may not be a major limitation in practice.

9.6   The three problems listed above may to some extent be overcome by comparative studies using a common method of case ascertainment in a large number of areas with different mixes of pollutants. A good example of this approach is the PAARC study,[1] in which adults and children were studied in 28 areas of

7 French cities with variable pollutant mixes. Unfortunately, no data on wheeze or asthma were presented. Two other examples, linking air pollution measurements to national health surveys in the United States, are also uninformative about wheezing illness, concentrating on lung function[2] or restriction of activity due to a wide and unspecified range of respiratory conditions.[3]

9.7 Early multi-centre studies of air pollution and child health in the UK concentrated on relationships with cough, phlegm and diagnoses of bronchitis or pneumonia, and published reports contain no information on wheezing or asthma.[4-8] In later studies conducted during the 1970s, wheeze and asthma were considered only as part of a symptom score derived from responses to six questions on respiratory symptoms and diagnoses.[9,10] Many of these studies were influential in early discussions about the dose-response relationships for long-term health effects of SO_2 and particulate air pollution.[11] It is, therefore, regrettable that asthma and wheezing were not specifically addressed in the published reports.

9.8 There are a few multi-centre prevalence comparisons relating pollution exposure to wheezing illness, although none have been conducted in the UK. These are collated and discussed in the next section and summarised in Table 9.1. Such an overview is an imperfect substitute for a single comparison based on upwards of 20 centres with variable pollutant mixes, but it can begin to overcome some of the methodological limitations of individual two-point comparisons. Studies have been included only if data are presented on the prevalence of wheeze or diagnosed asthma (the former being preferred). In reviewing the studies of adults, evidence from non-smokers has been emphasised because of the known association of wheeze with active smoking and the greater likelihood that wheeze has an asthmatic basis in non-smoking adults than in smokers.

Comparisons with measured pollutant levels

WHO collaborative studies in Europe

9.9 An early attempt at a multi-centre comparison of respiratory illness, lung function and urban smoke and SO_2 exposure among primary school children was coordinated by the WHO Regional Office for Europe in the 1970s.[12] Although this was conceived and analysed as a comparison across 20 areas, the fieldwork took place in 8 countries, each with a different language, and there was no cross-validation of responses to the translated symptom questionnaire. The results are, therefore, more usefully viewed as a set of within-country comparisons, each based on 2–4 areas. The symptom question common to all areas was: *"Does your child's chest ever sound wheezy or whistling?"*. In some countries, the prevalence of wheeze in the last year was also ascertained. Unfortunately, no data were collected on potential confounding variables such as parental smoking, cooking or heating fuels. The results of each prevalence study are summarised in Table 9.1, except for the Yugoslav component, where no air pollution data were obtained for the rural (low pollution) areas. A lower prevalence of wheeze was obtained for 1,990 children in urban Sarajevo (9.3%) than for 1,886 children in three surrounding rural areas (12.0%). The annual mean smoke and SO_2 levels in Sarajevo in 1975–76 were 131 $\mu g/m^3$ and 171 $\mu g/m^3$ (60 ppb), respectively.

WHO study: Eastern European centres[12]

9.10 Comparisons of polluted and unpolluted areas in Czechoslovakia, Romania and Poland were fairly consistent in finding a higher prevalence of wheezing (a 1.5–2-fold variation) in the more polluted towns. The exception was a higher prevalence of wheeze ever in the Polish town of Nowy Targ compared with more polluted Cracow. However, a comparison of wheeze in the past year among the Polish centres showed a different pattern, correlating more closely with the levels of pollution (11.1% in Cracow; 6.4% in Nowy Targ; 6.7% in Limanova). The Romanian comparison is particularly stark: annual mean levels of smoke and SO_2 in the coal-mining town of Petrosani were higher than those of east German cities in the 1980s, or London in the 1950s; whereas the control town of Brasov had levels of pollution similar to many areas of Britain today (see Chapter 7).

WHO study: Western European centres[12]

9.11   In contrast, polluted and unpolluted areas in Denmark, The Netherlands, Greece and Spain generally had a similar prevalence of wheeze. The exception was in Denmark, where the rural area (with very low pollutant levels) had a significantly lower lifetime prevalence of wheeze than the lightly polluted urban areas of Copenhagen and Odense. The prevalence of diagnosed asthma was similar in the three Danish centres (2.4% in Copenhagen; 2.3% in Odense; 2.3% in the rural area). A more detailed report of the Danish arm of the collaborative study questioned whether socioeconomic factors, rather than air pollution, might be responsible for the low prevalence in rural Denmark.[13]

Turin, Italy

9.12   Two more recent studies of Italian children in polluted and less polluted areas have been published. Each of these has controlled for the possible confounding effects of parental smoking and socioeconomic status. The prevalence of wheeze and asthma in 1980–81 was compared among 2,385 children living in central Turin, peripheral Turin and a nearby suburban area.[14] After adjustment for socioeconomic factors and parental smoking, there were no substantial differences in the prevalence of wheeze (see Table 9.1), but asthma was more commonly diagnosed in the suburban area (2.7%) than in central or peripheral Turin (1.2% and 1.3%, respectively). This difference is based on small numbers of suburban children and is not statistically significant. Annual mean SO_2 levels were about four times higher in central Turin than in the suburbs, whereas there was a less marked difference in particulate pollution. Gas cooking was almost universal in all areas.

Rome, Italy

9.13   In 1986–88, a similar comparison was made between children living in central Rome, the industrial area of Civitaveccia (80 km north of Rome) and three relatively unpolluted rural communities in the province of Viterbo (130 km north of Rome).[15] Parental smoking did not vary greatly between the three areas, but the authors suggested that its effect on the prevalence of asthma and/or wheeze differed with the level of ambient pollution. Children of smoking parents in Viterbo were more than twice as likely to be wheezy (7.5%) as children of non-smoking parents in the same area (3.4%). In the more polluted area of Rome and Civitaveccia, a smaller difference emerged between non-smoking families (8.8% v 7.8%). The authors suggested that this might reflect a saturation of relevant personal pollutant exposure in the more polluted area, but a closer examination of the statistical analysis suggests that this differential risk of passive smoking could easily have arisen by chance. Nevertheless, the study draws attention to the need to consider indoor exposure as part of the total pollutant burden of a community.

Tennessee, USA

9.14   An unusual opportunity to study the long-term health effects of exposure to nitrogen oxides arose when a TNT factory was reopened in Chattanooga, Tennessee during the Vietnam war.[16] Four areas were defined for a survey of school children in 1968–69: a high NO_2 area close to the TNT plant; a high particles area downwind of another industrial zone; and two control areas with low levels of particulate and NO_x pollution. A total of 987 children were included in studies of ventilatory function. The only information on asthma relates to the prevalence of diagnosed asthma ascertained by a screening questionnaire, and not to symptoms such as wheeze. Asthma was reported for 9% of children in the high particulate area, 4% in the high NO_2 area and 4% and 6% in the two control areas. With the small numbers studied in each area these differences are not statistically significant. Although this community was extensively investigated for other acute and chronic effects of NO_x pollution,[17–19] no further data relating specifically to wheezing illness has been published.

Arizona, USA

9.15   An unusual pattern of SO_2 pollution was a feature of smelter operations around the Arizona town of Morenci.[20] Changes in wind direction through the day resulted in brief hourly peaks of SO_2 exposure of over 800 $\mu g/m^3$ (280 ppb) both indoors and outdoors, against a background of less than 100 μg/m³ (35 ppb). Relatively small samples of children from four areas with widely differing exposures to the smelter plume were studied in the late 1970s.[20] There were no statistically significant differences in the prevalence of asthma or wheeze, although the prevalence of wheeze was greater among 201 children in the least polluted area (7.5%) compared to about 5% among 477 children living in areas affected to a varying degree by the smelter emissions (see Table 9.1 for details). Parental smoking and use of gas for cooking were present with similar frequency in each area.

Six Cities, USA

9.16 An extensive and influential investigation of respiratory health among children in six cities of eastern and mid-western USA was carried out in the late 1970s and early 1980s. The most detailed pollutant data relate to the period 1980–81, and cross-sectional prevalence data for this period are discussed here.[21] An earlier report focusing on particulate and SO_2 pollution will be discussed in a later section, as it contains information of relevance to changes in prevalence over time.[22] The Six Cities study enrolled children aged 6–9 in 1974–79 in selected urban areas for a longitudinal study of respiratory symptoms, lung function, and their relationship to outdoor and indoor environmental factors. In 1980–81, 5,422 children were re-examined at age 10–12 years.[21] At this time, there was a twofold variation in annual mean TSP levels across the six cities, and levels of PM_{15}, $PM_{2.5}$, and fine-fraction aerosol sulphate were also measured. There was also an eightfold variation in mean SO_2, a threefold variation in NO_2, and a twofold variation in ozone levels between centres (see Table 9.1). The least urbanised communities (Portage and Topeka) had the highest ozone levels, but the lowest levels of other pollutants. There was no obvious nor significant association of the prevalence of wheeze with level of any pollutant. The prevalence of doctor-diagnosed asthma was highest in Portage (5.1%) and Topeka (5.9%), intermediate in Kingston (4.4%) and lowest in Watertown (3.2%) and St Louis (3.4%) and Steubenville (3.3%). This ranking was similar to that for ozone levels, and the correlation between diagnosed asthma prevalence and ozone was statistically significant. Hay fever symptoms did not show the same trend, being most prevalent in St Louis (32.8%) and least prevalent in Watertown (12.1%).

Ontario, Canada

9.17 The possible long-term adverse effects of secondary pollutants (ozone and sulphates) were studied in a comparison of two rural Canadian communities in 1983–84.[23] The main source of pollution in southwestern Ontario and southern Manitoba is long-range transport of ozone, fine particles and aerosolised sulphate, rather than local emissions. Annual mean pollutant levels for the wider study are shown in Table 9.1. Special monitoring was carried out during winter 1983–84 in two agricultural service towns (Tillonsburg, Ontario and Portage la Prairie, Manitoba). Only the higher levels of sulphate differed significantly between the two centres, being higher in Tillonsburg (3.2 μg/m³) than Portage la Prairie (0.9 μg/m³). This detailed monitoring did not include summer measurements, and did not confirm whether a difference in ozone exposure existed between the two communities. Use of gas for cooking was more common in Tillonsburg (13% v 1%) but parental smoking habits were similar. A comparison of 1,317 children in the two communities found a similar prevalence of wheeze (2.7% in Tillonsburg, 2.4% in Portage la Prairie) and asthma (3.1 v 2.9%), but a significantly higher prevalence of inhalant allergies in Tillonsburg (11.0% v 5.0%). This finding is difficult to interpret because the pollutant exposure of the two communities does not appear to differ greatly and "inhalant allergy" is not defined in the published report.

California, USA

9.18 Two areas within Los Angeles county were selected for a detailed study of the effects of photochemical pollution on respiratory health of adults.[24] 2,369 residents of Glendora, an area with very high concentrations of oxidants and high levels of NO_2 and sulphates, were compared with 3,192 adults living in Lancaster, an area with low levels of most pollutants. There was more than a threefold difference in mean NO_2

levels, whereas the levels of total oxidants, hydrocarbons, particles, SO_2 and sulphates in Glendora were about twice those in Lancaster (see Table 9.1). The age-adjusted prevalence of wheezing reported at the entry examination was almost twice as high in Glendora as in Lancaster, among both smokers (25.5% v 15.3%) and lifelong non-smokers (14.3% v 7.8%). There was also an excess of "frequent chest illness" in Glendora, but the prevalence of diagnosed asthma, bronchitis or emphysema was similar in the two areas (10% v 11%). Later studies in these areas concentrated on measures of ventilatory function and do not offer data on asthmatic symptoms.[25,26]

California, USA

9.19   The possible adverse respiratory effects of photochemical oxidant pollution were also studied among Seventh Day Adventists in California in 1977. This group avoids cigarette smoking for religious reasons and a high proportion of the sample (78%) were lifelong non-smokers. The population of 4,379 current non-smokers who had lived for 11 years or more in a high pollution area (Los Angeles basin) were compared to 2,287 non-smokers from one of two low pollution areas (San Francisco or San Diego). Pollutant levels in the three areas are reported in terms of hourly exceedances of guideline levels, rather than annual means, although illustrative data for central Los Angeles and San Francisco during an earlier period (1969–72) are reported by Linn *et al*[27] (see Table 9.1). In the polluted Los Angeles basin hourly average concentrations of total oxidants, SO_2, sulphates and total particles frequently exceeded local guideline levels. Exceedances were uncommon in the other two areas during 1973–1976. The prevalence of "breathing ever wheezing or whistling" was 13.2% in the low pollution areas and 13.0% in the polluted area. The prevalence of doctor-diagnosed asthma was also similar (6.4% v 6.7%). This contrasts with the findings for lifelong non-smokers in Glendora and Lancaster.[24] There were small differences in the proportions of past smokers (23% v 19%) and occupational exposure to dust or fumes (7% to 5%) between the low and high pollution areas, respectively. These would tend to introduce a conservative bias, but it is unlikely to have been strong enough to obscure an epidemiologically important adverse effect of exposure to photochemical pollution. In subsequent analyses of the same religious group, lifetime exposure to a range of pollutants has been estimated from information on areas of residence in the past.[28,29] Unfortunately, these more detailed analyses do not specify asthma or wheezing as the outcome variable.

9.20   More recent publications from the Seventh Day Adventists cohort have related the incidence of asthma to estimates of lifetime exposure to oxidant air pollution and particles, occupational exposures and passive smoking.[30–32] Cumulative pollution exposures were derived from the locations of home and work for each individual, linked to pollutant concentrations at fixed site monitors over the past 20 years, using spatial interpolation to estimate mean monthly concentrations at each zipcode location. New cases of asthma between 1977 and 1987 were defined on the basis of physician diagnosis. There were 27 incident cases among 1,305 men, and 51 among 2,272 women. After adjustment for age, educational level and workplace tobacco smoke exposure, the risk of developing asthma increased significantly with cumulative particulate exposure (relative risk [RR] 1.74, 95% CI: 1.11–2.72, per 1,000 hours per year in excess of 200 μg/m³ TSP).[31] Cumulative ozone exposure was not significantly related to incident asthma in both sexes (RR 1.31, 95% CI: 0.96–1.78, per 10 ppb increase in mean annual exposure), although among men the risk was significantly elevated (RR 3.12, 95% CI: 1.61–5.85, per 10 ppb).[30] In the absence of a prior reason to analyse the sexes separately, this subgroup analysis should be interpreted with caution. Overall, this is a well designed, carefully conducted and thoroughly analysed study of chronic health effects which serves as a model to be replicated elsewhere. Unfortunately, estimated personal exposures to particles and ozone were highly correlated (r=0.72) so it is not possible to distinguish with confidence the independent effects of each pollutant. However, the balance of evidence is in favour of particles having the greater influence on asthma incidence.[32] This would be consistent with the adverse of environmental tobacco smoke exposure on asthma incidence in this non-smoking cohort (RR 1.50, 95% CI: 1.23–1.82, per 10 years worked with a smoker).[31]

Pennsylvania, USA

9.21   Cross-sectional studies of both adults[33] and children[34] were carried out in Chestnut Ridge, Pennsylvania in 1978–79. This is a hilly rural area containing four coal-fired electricity generating plants emitting SO_2 and particulate pollution. The area of residence of each subject was classified as high, medium or low pollution based on SO_2 and TSP levels monitored at 17 sites in the region. Among 5,557 women who had never smoked, the prevalence of wheeze on most days or nights increased significantly across the three categories of pollution, being almost twice as high in the most polluted areas (10.0%) as in the least polluted areas (5.1%).[33] The opposite trend was observed among children, although this inverse correlation with pollution level did not reach statistical significance.[34]

Connecticut, USA

9.22   A significantly *lower* prevalence of diagnosed asthma was found in Ansonia, a previously heavily polluted town in Connecticut, compared to a contrasting rural site, Lebanon.[35] Unfortunately, the methods and results of this paper (published in Nature) are poorly described and no prevalence data are presented for respiratory symptoms including wheeze. The difference in air pollution levels at the time of this survey was relatively small (see Table 9.1).

Israel

9.23   Children from two coastal communities in Israel were surveyed in 1980.[36] Mean levels of SO_2 in Ashdod, an industrial town, were four times higher than in Hadera. Nitrogen dioxide levels were twice as high in Ashdod as Hadera (see Table 9.1). There was a statistically higher prevalence of most asthmatic symptoms among 7–11 year old children in Ashdod; wheezing with a cold (17.0% v 14.0%), wheezing without a cold (9.8% v 7.1%), wheezing with shortness of breath (13.6% v 10.6%), a diagnosed asthma (13.0% v 9.8%). Adjustment for ethnic origin, domestic crowding and heating in the home made little difference to these results.

Singapore

9.24   A survey of 3,216 children aged 6–14 years living in industrial, urban and rural areas of Singapore was carried out in 1983.[37] Mean air pollutant levels are not reported, but a graphical presentation of monthly averages suggests that NO_x levels in the urban area were approximately double those in the rural and industrial areas. Sulphur dioxide levels were higher in the industrial area than in the urban area and rural areas. Particulate pollution was similar in all three areas (see Table 9.1). The prevalence of wheeze in the past year did not differ significantly between the three areas (see Table 9.1), nor was there any significant variation in the frequency of asthmatic attacks. However, there were ethnic and socioeconomic differences between the areas which were not taken into account in the analysis, and as the relationship of these variables to wheezing illness is not presented, no assessment of their possible confounding effect can be made.

Beijing, China

9.25   A carefully conducted study of 1,576 lifelong non-smokers in Beijing assessed the effect of both indoor and outdoor particulate pollution on respiratory symptoms in adults aged 40–69 years.[38] Respondents were classified by area of residence (industrial, residential, suburban) and by type of cooking fuel (coal or gas). Long-term outdoor particle levels measured over the preceding five years were high, especially in the industrial and residential areas. Special monitoring for short periods in the homes of respondents suggested generally lower levels of particles indoors than outdoors. Indoor particle levels were increased in the homes using coal for cooking or heating, but the ranking of the three geographical areas with respect to indoor exposure was the same as that for outdoor exposure to particles and SO_2. The prevalence of wheeze did not differ greatly between the three geographical areas (see Table 9.1), but there was a graded relationship with use of coal for heating or cooking. However, after adjustment for indoor sources, there was a significant twofold excess of wheeze in the industrial area, compared to the suburban area. There was also a significant trend of increasing prevalence of wheeze with increasing use of coal as a domestic fuel, with a 2.5-fold difference in prevalence of wheeze between respondents using coal for both cooking and heating, and those not using coal for either. When indoor and outdoor particle exposures were combined in a single index, this was significantly related to wheeze, independent of sex; age; education; annual income; indoor crowding; passive smoking; and occupational exposure to dusts, gases or fumes. This detailed and informative study in a community with persisting high levels of "old-fashioned" pollution thus suggests that both indoor and outdoor pollutant exposures affect respiratory health, but their specific relationship to asthma is difficult to assess, because of the relatively old age-group being studied, among whom other forms of obstructive airways disease may be common, even among lifelong non-smokers.

Exposure to point sources of pollution

9.26 A few studies have compared the prevalence of wheezing illness in areas around a point source of pollution, without quoting measured pollutant levels. Proximity to the source of pollution is presumed to indicate greater exposure, although in only one study[39] was this assessed with reference to pollution monitoring. These studies, which are not included in Table 9.1, are discussed below.

Haifa, Israel

9.27 An objective indicator of asthma (exercise-induced airflow reduction) was compared among 160 children living in an unpolluted urban area of Haifa, and 153 children living in a kibbutz in rural Haifa close to "gross" pollution from quarries, cement works, chemical industries and oil refineries.[40] The proportion of children with a reduction of 15% or more in peak expiratory flow rate after 6 minutes free running was 5.6% in the urban area and 11.8% in the rural area. However, after adjustment for differences in the age composition of the two samples, this difference was not statistically significant. It is difficult to interpret its epidemiological significance as no pollutant levels are quoted.

Norrköping, Sweden

9.28 Pollution from a pulp and paper mill was one of several outdoor and indoor exposures considered in a study of children in the Norrköping area of Sweden in 1985.[41] Particles, SO_2 and hydrogen sulphide emissions were of concern, but no recent pollutant concentrations are quoted. The prevalence of allergic asthma (defined as a breathing problem or wheezing when in contact with trees, grass, flowers or furred animals) was higher in the vicinity of the mill, although after adjustment for parental smoking, dampness and type of dwelling, the relative increase (1.3) was not statistically significant. There was a similar excess of allergic rhinitis around the pulp and paper mill (relative risk 1.3), which was statistically significant after adjustment for other factors. However, indoor environmental factors (parental smoking and dampness in the home) were of greater importance than area of residence in this study.

Alberta, Canada

9.29 Sulphur dioxide and hydrogen sulphide emissions from two natural gas refineries in rural Alberta were suspected by the local populations as a cause of excess illness in the area downwind of the refineries.[39] Measurements of total sulphation in ambient air over a number of years were used to define the "plume" of pollution, although concentrations of SO_2 and H_2S are not reported. When surveyed in 1985, the prevalence of wheeze among children aged 5–13 years was 20% among 113 children in the most highly sulphated area; 11% of 353 children in a surrounding less sulphated area; and 12% of 203 children in an unpolluted reference area. These differences are not statistically significant, but there was a statistically significant association of wheeze with sulphation across all measured levels of pollution. This trend was not seen for adults; among lifelong non-smokers the prevalence of wheeze was 14% in the highly polluted area, 18% in the less polluted area and 13% in the reference area. The implications of this study are uncertain because the nature and levels of pollution exposure are poorly defined. The study was conducted in response to local concerns and the authors raise the possibility that awareness of symptoms, particularly by parents of children, may have been greater in the vicinity of the refineries.

Table 9.1 Comparative studies of prevalence of wheezing illness in areas with different pollution levels

Author(s)	Year, Study area		Annual Means (μg/m³)					Symptom	Subjects	Prevalence	
			Smoke	TSP	SO₂	NO₂	O₃		Age	No	(%)
Colley[12]	1973, Czechoslovakia	Teplice/Most			147			Ever wheezy or	8–10	2170	16.0
		4 "low" towns			78			whistling		2106	8.7
Colley[12]	1974, Poland	Cracow	187		124			Ever wheezy or		1921	17.1
		Nowy Targ	82		57			whistling	8–10	1319	19.8[a]
		Limanova	53		41					565	11.7
Colley[12]	1973–5, Romania	Petrosani	353		162			Ever wheezy or	8–10	1141	23.5
		Brasov	37		9			whistling		1910	16.6
Colley[12]	1973, Netherlands	Westland (high)	29		148			Ever wheezy or	9–11	2198	9.2
		Westland (low)	9		48			whistling		276	11.6
Colley[12]	1973, Denmark	Copenhagen	29		69			Ever wheezy or		1811	17.3
		Odense	17		28			whistling	7–11	1142	16.8
		Rural area	7		9					645	12.2
Colley[12]	1975, Greece	Athens			106			Ever wheezy or	8–11	1054	12.2
		Vygon			53			whistling		948	12.7
Colley[12]	1975, Spain	Erandio			185					685	11.0
		Bilbao			138			Ever wheezy or	7–11	878	11.2
		Portugalete			127			whistling		890	11.5
		Guecho			100					768	11.9
Spinaci et	1980–81, Italy	Turin (Inner)		c150	c200			Current		719	3.8
al[14]		Turin (Outer)		c130	c150			wheezing	11	1481	3.8[b]
		Turin (Suburbs)		c110	c50					185	3.7
Forastiere	1986–88, Italy	Rome		c150	c70			Asthma		1137	8.3
et al[15]		Civitaveccia		c40	c38			and/or	7–11	955	8.9[c]
		3 rural areas			c3			wheeze		932	6.2
von Mutius	1989–91, Germany	Leipzig		*			39	Wheeze	9–11	1051	20.0
et al[54]		Munich					58	ever		5030	17.0
Shy et al[16,19]	1968–69, USA,	High NO₂		81		156		Asthma	Child	306	4.2
	Chattanooga, Tennessee	High TSP		99		103				192	9.0
		Control 1		72		118				264	3.8
		Control 2		62		81				225	5.9
Dodge et al[20]	1978–80, USA, Arizona	Stargo			103			Attacks of	Child	59	5.1
		San Manuel			48			shortness		134	4.5
		Morenci			14			of breath		284	5.3
		Kingman			<4			with wheeze		201	7.5
Dockery et	1980–81, USA, 6	Portage, WI	34	12	12	76		Wheeze apart		812	9.6
al[21]	Cities Study	Topeka, KS	63	10	24	61		from with		1213	11.4
		Watertown, MA	54	44	37	44		colds on	10–12	777	6.6
		Kingston, TN	64	27	29	51		most days or		531	10.6
		St Louis, MO	80	57	42	46		nights in		996	8.9
		Steubenville, OH	71	80	42	36		past year		1093	9.6
Stern et al[23]	1983–84, South Canada	Tillonsburg			3		27	Any wheeze	7–12	609	2.7
		Portage la Prairie			0		19			708	2.4
Bouhuys et	1973 USA,	Ansonia		63	14	18	88	Ever had	7–64	458	2.6
al[35]	Connecticut	Lebanon		40	11	55	85	asthma	males	1142	6.7
Detels et	1973–74 USA,	Glendora		High	66	226	Very high	Wheezing	25–29	1104	14.3
al[24]	California	Lancaster		Low	29	66	Low		never	1310	7.8
									smokers		
Linn et al[27]	1976, USA,	Los Angeles		135		131	140	Breathing	25+	4379	13.0
	California	San Francisco		47		65	40	ever wheezy	non–	2287	13.2[d]
		and San Diego						or whistling	smokers		
Schenker et	1978–79, USA,	High		80	99			Wheeze most	17–24	total	10.0
al[33]	Pennsylvania	Medium		73	66			days or	females	n=	7.1[e]
		Low		64	62			nights	never	5557	5.1
									smokers		
Goren and	1980, Israel	Ashdod			28	34		Wheezing	7–8 &	936	9.8[f]
Hellman[36]		Hadera			4	15		without colds	10–11	1245	7.1
Goh et al[37]	1983, Singapore	Industrial		c70	c40	c40		Wheeze in	6–14	1202	5.6
		Urban		c70	c20	c80		past year		993	7.5[g]
		Rural		c70	c10	c40				1021	5.8
Xu and	1986, China,	Industrial		449	128			Wheeze or	40–69	568	11.3
and Wang[38]	Beijing	Residential		389	57			whistling in	never	540	13.5[h]
		Suburban		261	18			the chest	smokers	468	12.7

[a] Current non–smokers only. Illustrative pollutant data taken from Linn *et al* related to 1976–72.

[b] Mean pollutant levels estimated from graphical presentation. Prevalences adjusted for sex, social status, active and passive smoking.

[c] Mean pollutant levels taken from the centre of quoted ranges of annual means for monitors in each area.

[d] For comparison of wheeze in past year, see text.

[e] Persistent wheeze was inversely related to SO_2 level in 4071 children in the three same areas.

[f] Annual mean pollutant levels calculated from monthly means.

[g] Pollutant levels estimated from geographical presentation in the published report.

[h] A significant positive correlation with air pollution emerged after adjustment for indoor sources.

* For data on SO_2 and particles see original paper.

Walsall, UK

9.30   A survey of school children in Walsall was carried out in 1989 in response to local concern that ammonia, isocyanate and formaldehyde emissions from a local foundry, and unspecified toxins from a hazardous waste disposal plant might have adverse long-term effects on health.[42] Formaldehyde levels measured in 1987 had risen to a peak of 300 μg/m³, although long-term average exposures were some 100 times lower than this. Other pollutant levels are not quoted. The prevalence of wheeze in the past year among 1,334 7-year-old children in Walsall (11.1%) was similar to the figure obtained by similar methods in 1,274 children of the same age in Southampton (11.9%). The prevalence of wheeze was not significantly related to proximity to the pollution sources within Walsall: if anything the prevalence was higher (12.9%) among children attending schools more than two miles from the waste plant or any foundry.

Western Australia

9.31   Respiratory symptoms, skin prick responses to common aeroallergens and bronchial hyperresponsiveness to inhaled histamine were measured among primary school children in Lake Munmorah, a coastal town near two power stations, and Nelson Bay, another coastal town free from major sources of pollution.[43] The prevalence of wheezing (assessed by various questions) and asthma was approximately twice as high in Lake Munmorah as in Nelson Bay. There was a significantly higher prevalence of bronchial hyperresponsiveness in Lake Munmorah, but no difference in the skin test reactivity. However, subsequent monitoring of SO_2 and NO_2 levels in the two centres demonstrated very low levels of each pollutant in both areas; annual mean SO_2 2.0 μg/m³ (0.7 ppb) in Lake Munmorah, and 0.3 μg/m³ (0.10 ppb) in Nelson Bay; annual mean NO_2 2.0 μg/m³ (1.1 ppb) in Lake Munmorah, and 0.4 μg/m³ (0.2 ppb) in Nelson Bay. It therefore seems likely that pollutants other than these, or factors unrelated to air pollution, were responsible for the differences in prevalence observed.

Studies of hospital admissions

9.32   Two studies relating measured or estimated air pollution exposure to hospital admission for asthma have been published. In Buffalo, Erie county, USA, four air pollution zones with mean TSP levels estimated as ranging from less than 80 μg/m³ to more than 135 μg/m³ were identified.[44] Hospital admission rates for childhood asthma, eczema and diabetes were calculated for each area and standardised for indices of socioeconomic status, derived from census data. Admission rates for asthma and eczema, but not diabetes, increased with increasing pollution level, although the statistical significance of these trends was not assessed. The authors suggested that as both asthma and eczema were allergic diseases, particulate air pollution might promote the development of allergy. An alternative explanation is that patients in the polluted area were more likely to attend or be referred to hospital for non-life-threatening diseases (including asthma and eczema, but not diabetes). In a case control study of asthma admissions in Perth, Western Australia, no association could be found between the risk of admission and exposure to ambient SO_2 in the range of 1 to 46 μg/m³ (0.35 to 16.1 ppb) (annual mean).[45] Again, variable patterns of referral potentially complicate the interpretations and could have obscured a true effect.

Comment

9.33   The pollutants which have been most often measured in geographical comparisons relating the prevalence of wheezing illness to air pollution are SO_2 and (somewhat less often) airborne particles. The evidence relating each of these to asthma and wheezing at the area level is inconsistent. Comparisons within eastern Europe (often at high levels of smoke and SO_2) generally support an association with lifetime prevalence of wheezing in children, whereas similar studies elsewhere (often at lower levels of SO_2 and particles) present a mixed picture. Only two studies (in Pennsylvania, USA and Beijing, China) address specifically the association of adult wheeze with airborne particles in non-smokers. Both found a positive relationship, at very different absolute levels of ambient pollution. Whether this is due to asthma, or to other forms of chronic obstructive airways disease, is unclear.

9.34   Fewer studies have addressed possible hazards due to "newer" pollutants (NO$_x$ and ozone). There have been no studies of this type in the UK or Europe, where the mix of photochemical pollution may be different from that encountered in North America. Two competent studies of wheeze among non-smoking Californian adults have suggested an association of symptoms with chronic exposure to photochemical pollutants. Neither of these studies were able to distinguish the effects of ozone exposure from those of correlated pollutants. The Glendora-Lancaster comparison is based on smaller numbers but probably greater differences in photochemical pollution exposure, and suggested a doubling in prevalence of wheeze in polluted Glendora. In early publications, the larger Seventh Day Adventists study found remarkably little difference in wheeze prevalence associated with an approximate doubling of particles, NO$_2$ and ozone. However, the incidence of new cases of asthma was significantly related to particle exposure, which was highly correlated with ozone. The Six Cities study suggests an association of ozone levels with diagnosed asthma, but not with persistent wheezing, in circumstances where ozone is negatively correlated with other pollutants, the opposite of the Californian situation. The relevance of these findings in the UK context is uncertain.

9.35   In general, NO$_2$ levels are correlated with SO$_2$ and particle concentrations in area comparisons and it is difficult to distinguish their independent effects. The rather limited evidence from Chattannooga, Tennessee suggests that particulate pollution may be more important than NO$_2$ in determining local variations in asthma prevalence.

Variations within the united Germany

Air pollution

9.36   An unusual opportunity to study the effects of air pollution on health arose from the reunification of Germany after a period of more than 40 years during which genetically similar populations in the east and west of the country had experienced very different levels and mixtures of air pollutants. Industrial regions of the former German Democratic Republic (GDR), particularly the highly polluted areas of Bitterfeld, Halle and Leipzig, and the cities of Erfurt and Weimar, experienced levels of winter particulate and SO$_2$ pollution which were typical of British cities in the early 1960s.

9.37   Winter mean levels in these cities were around 300 μg/m^3 (105 ppb) SO$_2$ and 120 μg/m^3 total suspended particles (TSP) in the late 1980s, compared to 50 μg/m^3 (17 ppb) SO$_2$ and 80 μg/m^3 TSP in Munich.[46] Outdoor nitrogen oxide levels tended to be higher in the West German cities, reflecting the much denser road traffic, but 80% of families in Leipzig, compared to about 40% in West Germany[47] used unvented gas appliances, a major source of indoor NO$_2$ pollution.

9.38   Following reunification, air pollution levels fell dramatically in cities of the former GDR. For instance, in Erfurt, annual mean SO$_2$ levels decreased from around 210 μg/m^3 (73 ppb) in 1988 and 1989 to 90 μg/m^3 (31 ppb) in 1990 and 62 μg/m^3 (22 ppb) in 1992. This compares with annual mean SO$_2$ levels of about 30 μg/m^3 (10 ppb) in Essen (part of the Ruhr area of West Germany). The annual mean level of TSP in Erfurt declined from 150 μg/m^3 in 1988 to 70 μg/m^3 in 1992, compared to 65 μg/m^3 in Essen.[48]

Asthma

9.39   Very few studies of the health effects of air pollution were conducted within the former GDR before reunification. The original reports (in German)[49-53] have recently been reviewed (in English) by Wichmann and Heinrich.[48] Four studies compared children from the Bitterfeld/Halle area with children from Berlin or rural areas elsewhere in the GDR. These are consistent in finding a higher prevalence of respiratory symptoms in the more polluted areas. The nature of the symptoms is not described in detail, and the magnitude of the difference is reported from only one (large) study, in which prevalence of respiratory symptoms was twice as high in Halle as in the control areas. Two studies which assessed ventilatory function found lower levels among the children in Bitterfeld. Lung function indices and symptoms apparently improved among 43 children from the industrialised areas who were taken on holiday for four weeks in cleaner air, suggesting that at least some of the adverse effects of the polluted environment were reversible in the short term.

9.40   Immediately after reunification, a number of research groups initiated comparative studies of the prevalence of respiratory and allergic diseases in West and East Germany. Three groups have reported results of fieldwork carried out in the former GDR during 1991 and 1992. In some instances, comparisons are made with results of West German studies carried out in the late 1980s, although similar survey methods have been applied.

9.41   In 1989–90, an extensive survey of respiratory and allergic disease was conducted among a sample of 5030 9–11 year old children in Munich. The same methods were applied to a sample of 1051 children of the same age in Leipzig in 1991.[54,55] The lifetime prevalence of wheezing was higher in Leipzig (20%) compared to Munich (17%), but the prevalence of doctor-diagnosed asthma or wheezy bronchitis was higher in Munich (9%) than in Leipzig (7%). Children in Leipzig were twice as likely to have received a diagnosis of bronchitis (without wheezing): 31% v 16%. Objective evidence of bronchial hyperresponsiveness, assessed by dry cold air challenge, was present with similar frequency in the two cities. Asthmatic children in Munich more often reported their attacks to be triggered by allergic factors (pets, grass or dust), whereas asthmatic children in Leipzig more often cited foggy weather as precipitant. Doctor-diagnosed hay fever was almost four times more common in Munich than in Leipzig, although no difference in the prevalence of itchy skin rashes or eczema was apparent between the two centres.

9.42   Similar results have been reported for 6-year-old children in Leipzig, Halle and Magdeburg (polluted areas of East Germany) and Borken (rural West Germany). Within both East and West Germany, the prevalence of frequent colds and cough without colds was more common in urban than rural areas, and higher in the east than the west.[47] Doctor-diagnosed asthma and hay fever were almost twice as common in Köln and Düsseldorf as in other areas, but doctor-diagnosed eczema was about twice as prevalent in the East German centres, where it did not vary with level of pollution. These results are more difficult to interpret than the Munich-Leipzig comparison, because differences in diagnostic labelling which could account for the findings and symptoms specific to asthma have not been presented.

9.43   Comparison of adults aged 20–44 years in Hamburg and Erfurt in 1991 formed part of the ongoing EC Respiratory Health Survey. Only the preliminary results from a postal questionnaire to over 4,500 persons in each city have been published.[56] Wheezing or whistling in the chest in the past year was significantly less prevalent in Erfurt (14%) than Hamburg (21%). Nocturnal breathlessness and cough were also less common in Erfurt, as were diagnosed asthma and current medication for asthma. Early results relating to objective measures of lung function in about 20% of the sample suggest similar levels of resting lung function and bronchial hyperresponsiveness as assessed by histamine challenge in the two cities.[57]

Allergy

9.44   In the study of adults mentioned above, the prevalence of nasal allergies and hay fever was also substantially lower in Erfurt (13%) than in Hamburg (23%),[54] but possible differences in the pattern of disease labelling between the two centres were not evaluated. However, in the comparison of children in Munich and Leipzig, the seasonality of nasal symptoms was assessed without reference to diagnostic terms. The prevalence of seasonal rhinitis in spring and summer, typical of pollen allergy, was substantially less common in Leipzig than in Munich, and more children in Munich reported allergic triggers for symptoms of rhinitis. This was consistent with the marked difference in prevalence of doctor-diagnosed hay fever (2.4% in Leipzig, 8.6% in Munich).[54,55]

9.45 More objective evidence of differences in the prevalence of allergic sensitisation in East Germany has emerged from several recent studies. The prevalence of positive skin prick test reactions to a standard batch of common aeroallergens was 37% among 4,419 children tested in Munich, compared to 17% among 1,829 children in Leipzig and Halle. The difference between cities was independent of family size, itself a strong correlate of skin prick positivity, with a reduced prevalence of positive responses among children from larger sibships.[55]

9.46 Similar results were obtained for young adults in Erfurt and Hamburg.[57] The prevalence of skin prick positivity to one or more allergens was 22% among 731 tested in Erfurt compared to 38% among 1,049 tested in Hamburg. In contrast, however, mean serum IgE levels were about 35% higher in Erfurt.

9.47 In a smaller sample of about 250 6-year-old children from each of four areas, the prevalence of skin prick positivity was higher in the urbanised West German areas of Duisberg and Essen (about 45%) compared to the rural West German area of Borken (27%) and the highly polluted East German area of Halle (29%).[47]

9.48 A more extensive comparison of total and allergen-specific serum IgE levels was made among 2,054 6-year-old children in four East German and three West German areas with a variety of pollution exposures.[58] This found significantly higher mean total IgE levels in the East German children, which reflected a shift in the entire distribution of total IgE, and not simply an extension of its upper tail. However, there was no significant difference in the proportions with one or more positive tests for allergen-specific IgE in the east (24%) and the west (21%). The prevalences of diagnosed hay fever and IgE specific to grass pollen were also similar in the east and west German centres. In both East and West Germany mean IgE levels were higher in the more polluted areas, among children from more crowded homes, and those with a history of parasitic infections. In West Germany, but not East Germany, IgE levels were higher among children exposed to sources of indoor air pollution (parental smoking or unvented gas appliances), although these effects were only of borderline statistical significance.

9.49 A further study of allergen-specific IgE among 900 school pupils from Leuna (East Germany) and Duisberg (West Germany)[59] found little difference in the prevalence of IgE specific to outdoor allergens, such as grass, birch and mugwort, but a higher prevalence of IgE specific to indoor allergens such as house dust mites (five times more common) and cat fur (three times more common) in Duisberg than in Leuna. In contrast, the only significant east-west differences among 6-year-old children in the study by Behrendt et al[58] were for the outdoor allergens birch and mugwort, the former being twice as common in West Germany and the latter twice as common in the east.

Comment

9.50 Comparisons of the prevalence of asthma and allergy in the united Germany have generated many unexpected findings and have prompted more questions than they have answered. Despite the need for rapid mobilisation of field studies in the year or two after reunification, a number of objective measures have been applied to large population samples and, in contrast to earlier geographical comparisons in other countries (reviewed in the next section), the focus has often been on asthma and allergy. Nevertheless, it should be noted that the fieldwork in each survey took place *after* measured levels of air pollutants in the East German cities had fallen substantially, and the comparisons are mainly between urban centres with different pollutant mixes, rather than between polluted and unpolluted areas.

9.51 The results of these east-west comparisons are fairly consistent in finding a lower prevalence of positive skin prick responses in the polluted areas of East Germany than in West German cities. One study suggested that variations in allergic sensitisation might also occur between urban and rural areas in West Germany. A similar urban excess of skin prick positivity has recently been reported for 10–12 year old children in urban and rural areas of Sundsvall in northern Sweden, where the prevalence of positive reactions is more than twice that in the Polish town of Konin, exposed to higher SO_2 levels but similar NO_2 levels to urban Sundsvall.[60]

9.52   In contrast, serum IgE levels seem to be consistently higher in east German areas than in the west. It is unclear whether this is related to air pollution or to a higher prevalence of parasitic infestation, which is known to raise IgE levels. There is currently little evidence that the higher levels of IgE in east Germany are due to higher levels of allergen-specific IgE, or that they result in a higher prevalence of allergic disease. Indeed, the balance of evidence suggests a lower prevalence of inhalant allergy (though not of eczema) in the former GDR.

9.53   Comparisons of symptoms related to asthma are complicated by the probable difference in prevalence of allergy between East and West Germany. There appears to be an excess of irritant symptoms (including "bronchitis") in the most heavily polluted areas of the former GDR, balanced by a reduced prevalence of "allergic asthma". Objective tests of bronchial hyperresponsiveness show little east-west difference in either adults or children. None of the studies published to date have assessed whether the residents of the East German cities have more asthmatic symptoms than those in West German centres, after allowing for the apparent reduction in prevalence of allergy in the east. The possible influence of confounding variables, in particular cigarette smoking by the adult population, have not been thoroughly assessed in publications to date.

9.54   Geographical comparisions in the united Germany may encapsulate the changes in pollutant exposure which have occurred in Britain over the past 30–40 years, as the pattern of air pollution has shifted from "winter smog" to "summer haze". However, there have been many cultural differences between East and West Germany which have not applied in an historical context in Britain. The results of the German comparisons therefore, need to be intepreted with caution when attempting to explain trends in asthma and allergy in the UK in the postwar period.

Exposure to vehicle emissions

Population surveys

9.55   A few epidemiological studies in Britain, Japan and Germany have used distance from major roads, or traffic density, as an indicator of exposure to vehicle exhausts. Mean NO_x concentrations decrease with distance from the kerbside,[61,62] although the decline beyond 20 metres is small. Levels of suspended particulate matter were also about 20% higher close to the roadside in one location in Tokyo.[61] However, a study of personal exposure to NO_x and particles in Japan found that *indoor* sources (cooking and heating fuels, and environmental tobacco smoke) were the most important influences, although a small decline in personal exposure with distance from major roads was still discernable.[63]

Tokyo, Japan

9.56   Three surveys of a total of 4,822 women aged 40–59 years in different parts of Tokyo were conducted in 1979, 1982 and 1983.[61] The prevalence of respiratory symptoms was related to distance of the home from major roads, after exclusion of recent arrivals in the area. Three-quarters of the respondents were lifetime non-smokers and smoking habits did not vary greatly with distance from the road. The prevalence of "chronic wheeze" (wheeze on most days or nights, or wheeze apart from with colds) was significantly greater in women living within 20m of a busy road in 1979 (9.8% v 4.2%) and 1982 (7.4% v 5.3%) but not 1983 (5.3% v 5.5%). Adjustment for age, smoking habit, duration of residence, education, occupation and type of home heating made little difference to these results. There was also an excess of chronic cough and phlegm among women living close to major roads, so it is unclear whether reporting of all symptoms differed with area of residence, perhaps because of awareness of the purpose of the enquiry.

Nikko-Imaichi, Japan

9.57 A second Japanese study[64] compared the prevalence of seasonal allergic rhinitis in the Nikko-Imaichi district, which includes forests of Japanese cedars, a major source of allergenic pollen in Japan. Old cedars are also planted along three inter-city highways running through the area. Cedar pollinosis was defined by the presence of rhinitis or conjunctivitis during the pollination period (March-April) as determined by a questionnaire survey of 3,133 respondents from 631 families. A statistically significant difference was reported between the prevalence of cedar pollinosis in five areas: 13.2% in areas close to tree-lined inter-city highways; 8.8% in city farming areas close to the cedar forests; 9.6% in city and farming areas distant from the forests; 5.1% within forest areas with little traffic; and 1.7% in mountainous areas above the tree line (there were less than 100 subjects in each of the latter two groups). Pollen counts were similar close to the forest and alongside the tree-lined highway, but lower in the city areas some distance from the forest. The authors suggested that both pollen exposure and local traffic density increased the likelihood of allergic sensitisation to cedar pollen. However, other differences between the respondents in each area were not explored and no quantitative estimate of exposure to traffic was obtained.

Bochum, Germany

9.58 Two German studies have used traffic density as a measure of exposure to vehicle exhausts. In Bochum, part of the Ruhr area, this information was obtained from 2,050 12–15 year old respondents to a respiratory questionnaire circulated in schools.[65] Streets of residence were categorised as a main road or side street, and by the reported frequency of heavy goods vehicle traffic on weekdays. Almost one-third of children reported frequent (23%) or constant (8%) truck traffic. The self-reported prevalence of wheezing in the past year increased significantly, and in a graded fashion, from 18% in streets without truck traffic to 28% in streets with constant truck traffic. There was also a significant trend in the self-reported prevalence of hay fever or allergic rhinitis in the past year, from 18% to 26%, respectively. These trends were not substantially affected by adjustment for a range of individual factors, including passive and active smoking, sibship size, household pets and home furnishings. The association of wheezing and allergic rhinitis with heavy goods traffic was stronger than with a general index of traffic density, although this may reflect the greater ease of recall and classification of heavy goods traffic.

Munich, Germany

9.59 Traffic census data from more than 600 streets were used to classify school catchment areas in Munich according to the street within the area with the highest volume of traffic.[66] These data were linked to information of respiratory symptoms and measurements of ventilatory function and bronchial responsiveness to cold air which had been obtained in a cross-sectional study of 9–11 year old children in the city in 1989–90. After exclusion of children who had moved house within the last 5 years, and those of non-German nationality, linked data were available for 4,678 subjects. Most symptoms, including doctor-diagnosed asthma and allergic rhinitis, were not significantly associated with traffic density, but the lifetime prevalence of "frequent wheezing" (wheezing or whistling in the chest several times) was significantly and positively related to the volume of traffic. After adjustment for a range of factors, including parental education, passive smoke exposure, domestic cooking and heating fuels, the relative increase in prevalence of frequent wheezing was about 8% per 25,000 cars per 24 hours, corresponding to a relative increase of 44% across the range of traffic densities observed (7,000 to 125,000 cars per 24 hours). This is similar to the relative difference between light and heavy truck traffic in the Bochum study.[65] No correlation was found between traffic density and bronchial responsiveness to cold air. Measures of resting lung function were negatively correlated with traffic density, but this may have been because more children in densely trafficked areas had colds at the time of examination.

UK studies

9.60 Two recently published studies from the UK have related asthma prevalence[67] or hospitalisation[68] to local traffic density. The self-reported prevalence of wheezing in the past year was significantly *lower* among 448 teenage children from areas close to the M23 and M25 motorways (20%), than among 962 teenagers from other parts of East Surrey (26%).[67] The prevalence of more severe forms of wheezing (6–8%) and of diagnosed asthma (15%) differed little between the two groups. These results run counter to the German findings but are difficult to interpret without more extensive information on overall levels of traffic exposure and related pollution experienced by the two groups.

9.61 In a case-control study of preschool children admitted to hospital in Birmingham, 715 children admitted with asthma were significantly more likely to live in an area with high traffic flow (more than 24,000 vehicles per day) than 736 children admitted for non-respiratory diseases or a community control group of similar size.[68] The effects of proximity (<200m) to major roads and high traffic flow were similar and multiplicative, each increasing the risk of hospital admission for asthma by 40–50% (comparing cases to community controls). A highly significant dose-response with local traffic density was observed for children living within 500m of a main road, but only comparing asthma cases to hospital controls (no such trend is evident for cases v community controls). An important limitation of this statistically powerful study is the lack of information on socioeconomic confounding. Although asthma prevalence does not vary greatly with socioeconomic status, utilisation of paediatric hospital services does, and inclusion of a hospital control group may have been insufficient to adjust for confounding influences. Similar studies in areas where local traffic density is not strongly correlated with social deprivation are required to clarify the interpretation.

Occupational studies

9.62 No statistically significant differences in respiratory symptoms were observed between 128 Boston policemen heavily exposed to traffic, and 140 policemen posted at an "outskirt station" where 87% had never worked outdoors in traffic.[69] The prevalence of "wheezing on most days or nights" was slightly higher at the outskirt station (13.6% v 10.9%).

9.63 Among heavily exposed road tunnel workers,[70] the prevalence of wheezing did not differ significantly between 54 men employed for less than 7 years (20%) and 120 employed for more than 7 years (26%). The prevalence of doctor-diagnosed asthma was also similar, though based on very few cases (5% v 7%). Another study compared 190 toll bridge workers with 160 road tunnel workers, the latter being more heavily exposed to exhausts.[71] The prevalence of wheezing was greater in the tunnel workers (22%) than the bridge workers (13%). This difference, which was of borderline statistical significance, appeared to be concentrated among the non-smokers.

9.64 Two studies have specifically investigated occupational exposure to diesel fumes. Unfortunately, neither address directly the chronic effects on asthma or allergy. Work-related respiratory symptoms (including wheeze) were more commonly reported by workers in a diesel bus garage than in a comparison population of battery workers,[72] but chronic health effects were not evaluated. No data on symptoms specific to asthma were presented in a study of long-term effects of exposure to diesel emissions in coal mines.[73]

Comment

9.65 The strength of the relationships between asthma prevalence or hospitalisation and proximity to traffic which have been reported in population surveys is remarkable, given the crude nature of this assessment of exposure to vehicle exhausts. This evidence is difficult to reconcile with the lack of urban-rural variation in wheezing illness (see Chapter 7), and the absence of a marked excess of asthma or allergic disease among the occupational studies of workers heavily exposed to traffic. Until local traffic density has been more thoroughly validated as an indicator of personal exposure to specific pollutants of interest, it will remain difficult to place a causal interpretation on these findings.

**Changes in prevalence
related to long-term
changes in pollution**

Increasing levels of pollution

9.66   An increase in the prevalence of asthma following an abrupt increase of
pollutant exposure of a population would argue in favour of a causal relationship.
Perhaps the clearest examples of such an association arises from studies of
occupational asthma due to organic pollutants. There are only two examples of an
apparent epidemic of asthma induced by population exposure to inorganic pollution,
and in neither of these was the clinical disease entirely typical.

9.67   Many cases of a form of asthmatic bronchitis occurred among previously
healthy US servicemen and their families who were stationed in the Tokyo-
Yokohama area of Japan during the 1950s.[74] This area, the Kanto plain, was highly
industrialised and surrounded by hills. Autumn and winter smogs, often developing
in the afternoon and evening, were common, although the constituent pollutants
were not characterised or measured precisely. The clinical picture was of cough,
wheeze and breathlessness similar to bronchial asthma, which was exacerbated by
the smog and relieved by travel outside the Kanto plain. In contrast to the usual form
of asthma, conventional bronchodilator treatment was ineffective. Some patients
became symptom-free on return to the United States, but others appeared to develop
a progressive form of airways obstruction. The local Japanese residents did not
appear to suffer from this complaint.[75] Subsequently, case-control studies
demonstrated an excess of cigarette smoking among servicemen who developed
"Tokyo-Yokohama asthma" (97%, compared to 75% of controls).[76] There were also
increased levels of circulating eosinophils in the patients and all sufferers had at least
one positive skin test reaction to common aeroallergens. This suggested that an
allergic mechanism might be responsible for the unusual features of Tokyo-
Yokohama asthma.[76] The nature of this epidemic and its relevance, if any, to the
more common presentations of asthma remain uncertain.

9.68   In another Japanese city, Yokkaichi, an atypical form of adult asthma emerged
among the local residents following rapid industrialisation with oil and
petrochemical refineries in the 1950s and 1960s.[77] Symptoms were at first similar to
asthma, often with severe attacks, progressing after one or two years to chronic cough
and phlegm. Males and smokers were at greater risk and few patients were under the
age of 50. In contrast to other forms of adult asthma in the area, few patients had
positive skin reactions to local aeroallergens. Patients experienced considerable
symptomatic improvement when they moved to non-polluted districts, and there was
a graded geographical association between disease prevalence and the ambient level
of SO_2 pollution across the city. There were also temporal variations in the asthma
attack rate which often corresponded to fluctuations in measured SO_2 levels. It has
since been argued that a possible cause of this epidemic was concentrated sulphuric
acid mists emitted by a titanium oxide plant on the windward side of the main
residential area.[78] As a result of successful legal action by local inhabitants against six
industrial companies, emission controls were introduced and levels of sulphur oxides
in the area improved markedly from 1972 onwards, falling by about one half between
1974 and 1982. Asthma mortality rates in the over sixties declined from a peak of 820
per million during 1967–1970 to 600 per million in 1979–1982.[79] There was also a
decline in asthma mortality among children and young adults, although this is based
on very small numbers of deaths and these groups did not appear to suffer greatly
from the atypical form of asthma in Yokkaichi.[77]

Decreasing levels of pollution

9.69   Two UK studies of air pollution and respiratory disease were repeated after an
interval during which smoke and SO_2 levels had declined as a result of clean air
legislation.[8,10] Unfortunately, neither of these report information specific to asthma
or wheezing.

9.70   An unusual pattern of year-to-year fluctuations in SO_2 and particulate
pollution occurred in the Japanese town of Awara-machi, set in a predominantly
rural area with no major sources of air pollution except two power stations which
were commissioned in 1973 and 1978.[80] Pollution levels rose from 1971 to 1973 as a

result of the first station operating, then declined from 1974 to 1977 as a result of statutory controls on emissions, increasing transiently in 1978 due to the opening of the second power station. However, the annual fluctuations in SO_2 and particle concentrations in central Awara were not large (SO_2 ranging from 40 to 49 $\mu g/m^3$ (14 to 17 ppb) and TSP from 30 to 40 $\mu g/m^3$ during 1971–79). Nitrogen dioxide levels were measured from 1974 onwards and declined during 1974–78, increasing slightly in 1979. Mass medical examinations of about 1,500 school children were carried out each summer, including skin prick tests with common local aeroallergens to determine the presence of atopy. The prevalence of "wheezing with colds" was analysed each year among atopic and non-atopic children separately in relation to the pollution levels for the previous year. No significant correlation was found in either group for SO_2 or particles, but there was a strong direct correlation between the prevalence of wheeze among atopic subjects and NO_2 levels during 1974–79 (only). This was a period during which both prevalence and NO_2 levels were falling steadily and the correlation probably reflects these coincident trends, rather than a causal relationship.

9.71 In the US Six Cities study, the prevalence of symptoms among school children was assessed on four occasions at yearly intervals in each centre during 1974–79.[22] An analysis of changes in prevalence with changes in pollution levels over these four years within each centre suggested a statistically significant *inverse* relationship between the prevalence of wheeze and annual mean levels of TSP and SO_2 for the preceding year. This association, the opposite of what might have been expected from an adverse effect of pollution, was largely attributable to the findings in two cities: St Louis and Steubenville, where an increasing prevalence of wheeze occurred at the same time as large declines in TSP and SO_2 levels. It is unlikely that this represents cause and effect, but the results argue against a medium-term decline in the burden of childhood asthma following modest improvements in air quality.

9.72 Changes in the prevalence of respiratory symptoms among adults in Berlin, New Hampshire, USA were studied over the period 1961–1973.[81,82] Over this period particulate pollution declined (annual mean TSP levels: 180 $\mu g/m^3$ in 1961; 131 $\mu g/m^3$ in 1966–67; and 80 $\mu g/m^3$ in 1973). Ambient SO_2 levels were similar in 1961 and 1973 (about 66 $\mu g/m^3$, 23 ppb, annual mean), although they were slightly lower in 1966–67. Among 274 lifelong non-smoking women aged 25–74, the prevalence of wheezing declined by a small amount between 1961 and 1967 (2.5% v 2.3%). Among 364 female non-smokers studied in 1967 and 1973, there was a further small decline in prevalence (2.5% v 2.2%). Fewer men who had never smoked were included in the study, but the trends in prevalence of wheezing were consistent with those for women. These results suggest a modest improvement in respiratory health of a community at the time that particle concentrations were falling. However, other changes may have been responsible for this decrease in prevalence. The prevalences are based on small numbers of subjects and the changes could have arisen by chance alone.

Comment

9.73 The epidemics of asthma encountered in association with increased pollution levels in Yokohama and Yokkaichi were unusual and are probably of marginal relevance to the burden of wheezing illness in Britain today. There is scant and inconsistent evidence relating changes in prevalence to improvements in air quality. The only statistically significant relationship was from the Six Cities study, where the prevalence of wheeze increased at a time of declining particulate and SO_2 pollution. The effects on asthma of the marked reductions in urban smoke and SO_2 pollution which have occurred in Britain over the past 30–40 years are therefore uncertain.

Conclusions

9.74 A number of geographical comparisons of high and low pollution areas have been reviewed. In many of these studies, asthma was not the focus and the published evidence is often less than satisfactory. In relation to "old-fashioned" smoke and sulphur dioxide pollution, the studies are inconsistent and most of those from

Western Europe and America have not found substantial differences in asthma prevalence between areas with different levels of smoke and sulphur dioxide pollution. Evidence suggesting an association of photochemical pollution exposure with development of asthmatic symptoms related mainly to non-smoking Californian adults. These have been well conducted studies but have not been able to reliably distinguish the effects of particle and ozone exposure. The type of particles involved are likely to be different to those encountered in the UK. However, these studies do not rule out an effect of chronic ozone exposure on the development of asthma in any age group.

9.75   The available evidence is inadequate for distinguishing the effects of long-term exposure to different pollutant mixes or for estimating the nature of the dose-response relationship to any measured pollutant. There is a need for multi-centre prevalence studies which specifically measure indicators of asthma and allergy in many areas with different pollutant mixes and concentrations. The comparative studies of East and West German cities and the Six Cities study in the USA have shown the way in terms of field protocols, but are deficient in their overall design. Upwards of twenty centres are required to generate the statistical power required to reliably quantify dose-response relationships and to distinguish the independent effects of different pollutants.

9.76   Although it is not possible to exclude a relatively weak relationship between outdoor air pollution exposure and the prevalence of asthma, it is apparent from several studies that indoor exposures are often a more important determinant of personal exposure to measured pollutants than ambient levels outdoors. This is not surprising in Westernised cultures where the majority of time is spent inside buildings. Opportunities exist for the effects of some of the pollutants of interest to be studied more efficiently by comparisons based on domestic or industrial exposure. Such investigations would include the possible relationship of asthma and allergy to environmental tobacco smoke (a source of particles and nitrogen oxides), gas cooking (nitrogen oxides) and occupational exposure to diesel fumes.

9.77   There are few examples of asthma emerging as a public health problem in association with new sources of pollution. When it has done, as in Yokohama and Yokkaichi, the clinical characteristics are atypical and the pollutants involved may also have been of an unusual nature. There is no convincing evidence of a decline in asthma prevalence in association with improvement in ambient air quality. Indeed, in the USA, as must have occurred in Britain, the prevalence of wheezing in children increased at a time when smoke and sulphur dioxide levels were falling. This could be due to the more powerful influence of other pollutants whose levels were increasing, but in the UK situation concentrations of both the likely candidates (nitrogen oxides and ozone) have been falling since measurements were started in the 1970s. It is thus unlikely that changes in outdoor air pollution exposure have been an important influence on trends in asthma prevalence in Britain.

9.78   The evidence from population studies cannot totally exclude the possibility that increasing levels of some unmeasured pollutant (such as diesel exhaust) might have contributed to changes in the prevalence of wheezing illness. It is notable that four of five studies which have examined vehicular traffic as an index of exposure have shown an association with symptoms of asthma. More studies of this nature are required, in combination with evidence of the degree of variation in personal exposure to vehicle related pollutants among residents of areas with different traffic densities.

References

1. PAARC Co-operative Group. Atmospheric pollution and chronic or recurrent respiratory diseases. I: Methods and materials. II: Results and discussion. Bull Physiopath Respir 1982; 18:87–116.

2. Schwartz J. Lung function and chronic exposure to air pollution: a cross-sectional analysis of NHANES II. Environ Res 1989; 50:309–321.

3. Ostro BD, Rothschild S. Air pollution and acute respiratory morbidity: an observational study of multiple pollutants. Environ Res 1989; 50:238–247.

4. Colley JRT, Reid DD. The urban and social origins of chronic bronchitis in England and Wales. BMJ 1970; 2:213–217.

5. Holland WW, Halil T, Bennett AE, Elliott A. Factors influencing the onset of chronic respiratory disease. BMJ 1969; 2:205–208.

6. Douglas JWB, Waller RW. Air pollution and respiratory infection in children. Br J Prev Soc Med 1966; 20:1–8.

7. Lunn JE, Knowelden J, Handyside AJ. Patterns of respiratory illness in Sheffield infant schoolchildren. Br J Prev Soc Med 1967; 21:7–16.

8. Lunn JE, Knowelden J, Roe JW. Patterns of respiratory illness in Sheffield junior schoolchildren. Br J Prev Soc Med 1970; 24:223–226.

9. Melia RJ, Florey CD, Swan AV. Respiratory illness in British schoolchildren and atmospheric smoke and sulphur dioxide 1973–7. I: Cross-sectional findings. J Epidemiol Community Health 1981; 35:161–167.

10. Melia RJ, Florey CD, Chinn S. Respiratory illness in British schoolchildren and atmospheric smoke and sulphur dioxide 1973–7. II: Longitudinal findings. J Epidemiol Community Health 1981; 35:168–173.

11. Department of Health. Advisory Group on the Medical Aspects of Air Pollution Episodes. Second Report: Sulphur Dioxide, Acid Aerosols and Particulates. London: HMSO, 1992.

12. Colley JRT, Brasser LJ. Chronic respiratory disease in children in relation to air pollution: report on a WHO study. Copenhagen: WHO, 1980.

13. Holma B, Kjaer G, Stokholm J. Air pollution, hygiene and health of Danish schoolchildren. Sci Total Environ 1979; 12:251–286.

14. Spinaci S, Arossa W, Bugiani M, Natale P, Bucca C, de Candussio G. The effects of air pollution on the respiratory health of children: a cross-sectional study. Pediatr Pulmonol 1985; 1:262–266.

15. Forastiere F, Corbo GM, Michelozzi P, Pistelli R, Agabiti N, Brancato G, Ciappi G, Perucci CA. Effects of environment and passive smoking on the respiratory health of children. Int J Epidemiol 1992; 21:66–73.

16. Shy CM, Creason JP, Pearlman ME, McClain KE, Benson BF, Young MM. The Chattanooga school children study: effects of community exposure to nitrogen dioxide. I: Methods, description of pollutant exposure and results of ventilatory function testing. JAPCA 1970; 20:539–545.

17. Pearlman ME, Finklea JF, Creason JP, Shy CM, Young MM, Horton RJM. Nitrogen dioxide and lower respiratory illness. Pediatrics 1971; 47:391–398.

18. Love GJ, Lan SP, Shy CM, Riggan WB. Acute respiratory illness in families exposed to nitrogen dioxide ambient air pollution in Chattanooga, Tennessee. Arch Environ Health 1982; 37:75–80.

19. Shy CM, Creason JP, Pearlman ME, McClain KE, Benson BF, Young MM. The Chattanooga school children study: effects of community exposure to nitrogen dioxide: incidence of acute respiratory illness. JAPCA 1970; 20:582–588.

20. Dodge R, Solomon R, Moyers J, Hayes C. A longitudinal study of children exposed to sulfur oxides. Am J Epidemiol 1985; 121:720–736.

21. Dockery DW, Speizer FE, Stram DO, Ware JH, Spengler JD, Ferris BG. Effects of inhalable particles on respiratory health of children. Am Rev Respir Dis 1989;139:587–594.

22. Ware JH, Ferris BG, Dockery DW, Spengler JD, Stram DO, Speizer FE. Effects of ambient sulfur oxides and suspended particles on respiratory health of preadolescent children. Am Rev Respir Dis 1986; 133:834–842.

23. Stern B, Jones L, Raizenne M, Burnett R, Meranger JC, Franklin CA. Respiratory health effects associated with ambient sulfates and ozone in two rural Canadian communities. Environ Res 1989; 49:20–39.

24. Detels R, Sayre JW, Coulson AH, Rokaw SN, Massey FJ, Tashkin DP, Wu MM. The UCLA population studies of chronic obstructive disease. IV: respiratory effect of long-term exposure to photochemical oxidants, nitrogen dioxide and sulfates on current and never smokers. Am Rev Respir Dis 1981; 124:873–880.

25. Detels R, Tashkin DP, Sayre JW, Rokaw SN, Coulson AH, Massey FJ, Wegman DH. The UCLA population studies of chronic obstructive respiratory disease. IX: Lung function changes associated with chronic exposure to photochemical oxidants: a cohort study among never smokers. Chest 1987; 92:594–603.

26. Detels R, Tashkin DP, Sayre JW, Rokaw SN, Massey FJ, Coulson AH, Wegman DH. The UCLA population studies of chronic obstructive diseases. X. A cohort study of changes in respiratory function associated with chronic exposure to SO_2, NO_x and hydrocarbons. Am J Public Health 1991; 81:350–359.

27. Linn WS, Hackney JD, Pederson EE, Breisacher P, Patterson JV, Mulry A, Coyle JF. Respiratory function and symptoms in urban office workers in relation to oxidant air pollution exposure. Am Rev Respir Dis 1976; 114:477–483.

28. Euler GL, Abbey DE, Magie AR, Hodgkin JE. Chronic obstructive pulmonary disease symptom effects of long-term cumulative exposure to ambient levels of total suspended particulates and sulfur dioxide in California Seventh-Day Adventist residents. Arch Environ Health 1987; 42:214–222.

29. Euler GL, Hodgkin JE, Abbey DE, Magie AR. Chronic obstructive pulmonary disease symptom effects of long-term cumulative exposure to ambient levels of total oxidants and nitrogen dioxide in California Seventh-Day Adventist residents. Arch Environ Health 1988; 43:279–285.

30. Greer JR, Abbey DE, Burchette RJ. Asthma related to occupational and ambient air pollutants in nonsmokers. J Occup Med 1993; 9:909–915.

31. Abbey DE, Petersen F, Mills PK, Beeson WL. Long-term ambient concentrations of total suspended particulates, ozone and sulfur dioxide and respiratory symptoms in a non-smoking population. Arch Environ Health 1993; 48:33–47.

32. Abbey DE, Lebowitz MD, Mills PK, Petersen FF, Beeson WL, Burchette RJ. Long-term ambient concentrations of particulates and oxidants and development of chronic disease in a cohort of non-smoking California residents. Inhalation Toxicol 1995; 7:19–34.

33. Schenker MB, Speizer FE, Samet JM, Gruhl J, Batterman S. Health effects of air pollution due to coal combustion in the Chestnut Ridge region of Pennsylvania: results of a cross-sectional analysis in adults. Arch Environ Health 1983; 38:325–330.

34. Schenker MB, Vedal S, Batterman S, Samet J, Speizer FE. Health effects of air pollution due to coal combustion in the Chestnut Ridge region of Pennsylvania: cross-sectional survey of children. Arch Environ Health 1986; 41:104–108.

35. Bouhuys A, Beck GJ, Schoenberg JB. Do present levels of air pollution outdoors affect respiratory health? Nature 1978; 276:467–470.

36. Goren AI, Hellmann S. Prevalence of respiratory symptoms and diseases in schoolchildren living in a polluted and in a low polluted area in Israel. Environ Res 1988; 45:28–37.

37. Goh KT, Lun KC, Chong YM, Ong TC, Tan JL, Chay SO. Prevalence of respiratory illness of school children in the industrial, urban and rural areas of Singapore. Tropical Geographical Med 1986; 38:344–350.

38. Xu X, Wang L. Association of indoor and outdoor particulate level with chronic respiratory illness. Am Rev Respir Dis 1993; 148:1516–1522.

39. Dales RE, Spitzer WO, Suissa S, Schecter MT, Tousignant P, Steinmetz N. Respiratory health of a population living downwind from natural gas refineries. Am Rev Respir Dis 1989; 139:595–600.

40. Colin A, Said E, Winter ST. Exercise-induced asthma in schoolchildren. A pilot study in two Haifa districts. Israel J Med Sci 1985; 21:40–43.

41. Andrae S, Axelson O, Björksten B, Fredriksson M, Kjellman NIM. Symptoms of bronchial hyperreactivity and asthma in relation to environmental factors. Arch Dis Child 1988; 63:473–478.

42. Symington P, Coggon D, Holgate S. Respiratory symptoms in children at schools near a foundry. Br J Ind Med 1991; 48:588–591.

43. Henry RL, Abramson R, Adler JA, Wlodarczyk J, Hensley MJ. Asthma in the vicinity of power stations. I: A prevalence study. Pediatr Pulmonol 1991; 11:127–133.

44. Sultz HA, Feldman JG, Schlesinger ER, Mosher WE. An effect of continued exposure to air pollution on the incidence of chronic childhood allergic disease. Am J Public Health 1970; 60:891–900.

45. Hunt TB, Holman CDJ. Asthma hospitalisation in relation to sulphur dioxide atmospheric contamination in the Kwinana industrial area of Western Australia. Community Health Studies 1987; 11:197–201.

46. Magnussen H, Jörres R, Nowak D. Effect of air pollution on the prevalence of asthma and allergy: lessons from the German reunification. Thorax 1993; 48:879–881.

47. Schlipköter HW, Krämer U, Behrendt H, Dolgner R, Winkler RS, Ring J, Willer H. Impact of air pollution on children's health. Results from Saxony-Anhalt and Saxony as compared to Northrhine-Westphalia. In: Proceedings of the 9th World Management Clean Air Congress, Montreal, Canada. Pittsburgh: Air and Waste Management Association, 1992; A-03.

48. Wichmann HE, Heinrich J. Health effects of high level exposure to traditional pollutants in East Germany. Review and ongoing research. Environ Health Perspect 1995; 103:29–35.

49. Thiemann HH, Weingartner L, Bromme W, Vorwald U. On the epidemiology of recurrent and chronic bronchopulmonary diseases—an examination of about 18000 children. I. Planning and organization of the investigation. Zeitschrift fur Erkrankungen der Atmungsorgane 1976; 145:98–103.

50. Thiemann HH, Bromme W, Vorwald U, Haidar H, Schmeckebier S. On the epidemiology of recurrent and chronic bronchopulmonary diseases—an examination of about 18000 children. 2. Investigations in the district of Quedlinburg. Zeitschrift fur Erkrankungen der Atmungsorgane 1979; 152:170–179.

51. Thiemann HH, Vorwald U, Bromme W, Weingartner L. On the epidemiology of recurrent and chronic bronchopulmonary diseases—an examination of about 18000 children. 6th report: summary of the investigation in the four districts. Zeitschrift fur Erkrankungen der Atmungsorgane 1979; 152:170–179.

52. Vorwald U, Thiemann HH, Bromme W, Gross G, Mucke B. On the epidemiology of recurrent and chronic bronchopulmonary diseases—an examination of about 18000 children. 4. Investigation in the cities of Halle and Halle-Neustadt. Zeitschrift fur Erkrankungen der Atmungsorgane 1976; 145:116–120.

53. Vorwald U, Thiemann HH, Bromme W, Klotz F. On the epidemiology of recurrent and chronic bronchopulmonary diseases—an examination of about 18000 children. 5. Investigations in Wolfen and Bitterfeld. Zeitschrift fur Erkrankungen der Atmungsorgane 1976; 145:121–125.

54. von Mutius E, Fritzsch C, Weiland SK, Röll G, Magnussen H. Prevalence of asthma and allergic disorders among children in the United Germany: a descriptive comparison. BMJ 1992; 305:1395–1399.

55. von Mutius E, Martinez FD, Fritzsch C, Nicolai T, Reitman P, Thiemann HH. Skin test reactivity and number of siblings. BMJ 1994; 308:692–695.

56. Nowak D, Heinrich J, Beck E, Willenbrock U, Jörres R, Claussen M, Berger I, Wichmann HE, Magnussen H. Differences in respiratory symptoms between two cities in Western and Eastern Germany: the first report in adults [Abstract]. Am Rev Respir Dis 1993; 147:A378.

57. Heinrich J, Nowak D, Beck E, Jörres R, Wassmer G, Boczor S, Claussen M, Berger J, Wichmann HE, Magnussen H. Comparison of bronchial responsiveness and atopy in two urban adult populations in Eastern and Western Germany [Abstract]. Am J Respir Crit Care Med 1994; 149(4 Part 2):A915.

58. Behrendt H, Krämer U, Dolgner R, Hinrichs J, Willer H, Hagenbeck H, Schlipköter HW. Elevated levels of total serum IgE in East German children: atopy, parasites or pollutants? Allergol J 1993; 2:31–40.

59. Klein K, Dathe R, Göllnitz S, Jäger L. Allergies—a comparison between two vocational schools in East and West Germany. Allergy 1992; 47(Suppl 12):259.

60. Bräbäck L, Breborowicz A, Dreborg S, Knutsson A, Pieklik H, Björksten B. Atopic sensitisation and respiratory symptoms among Polish and Swedish schoolchildren. Clin Exp Allergy 1994; 24:826–835.

61. Nitta H, Sato T, Nakai S, Maeda K, Aoko S, Ono M. Respiratory health associated with exposure to automobile exhaust. I. Results of cross-sectional studies in 1979, 1982 and 1983. Arch Environ Health 1993; 48:53–58.

62. Bower JS, Lampert JE, Stevenson KJ, Atkins DHF, Law DV. A diffusion tube survey of NO_2 levels in urban areas of the UK. Atmos Environ 1991; 25B:255–265.

63. Ono M, Hirano S, Murakami M, Nitta H, Hakai S, Maeda K. Measurements of particle and NO_2 concentrations in homes along the major arterial roads in Tokyo. J Japan Soc Air Pollut 1989; 24:90–99.

64. Ishizaki T, Koizumi K, Ikemori R, Ishiyama Y, Kushibiki E. Studies of prevalence of Japanese cedar pollinosis among the residents in a densely cultivated area. Ann Allergy 1987; 58:265–270.

65. Weiland SK, Mundt KA, Rückmann A, Keil U. Self-reported wheezing and allergic rhinitis in children and traffic density on street of residence. Ann Epidemiol 1994; 4:79–83.

66. Wjst M, Reitman P, Dold S, Wulff A, Nicolai T, von Loeffelholz-Colberg E, von Mutius E. Road traffic and adverse effects on respiratory health in children. BMJ 1993; 307:596–600.

67. Waldron G, Pottle B, Dod J. Asthma and the motorways—one district's experience. J Public Health Med 1995; 17:85–89.

68. Edwards J, Walters S, Griffiths RK. Hospital admissions for asthma in preschool children: relationship to major roads in Birmingham, United Kingdom. Arch Environ Health 1994; 49:223–227.

69. Speizer FE, Ferris BG. Exposure to automobile exhaust. I: Prevalence of respiratory symptoms and diseases. Arch Environ Health 1973; 26:313–318.

70. Tollerud DJ, Weiss ST, Elting E, Speizer FE, Ferris BG. The health effects of automobile exhaust. VI: Relationship of respiratory symptoms and pulmonary function in tunnel and turnpike workers. Arch Environ Health 1983; 38:334–340.

71. Ayres SM, Evans R, Licht D, Griesbach J, Reimold F, Ferrand EF, Criscitiello A. Health effects of exposure to high concentrations of automotive emissions. Studies in bridge and tunnel workers in New York city. Arch Environ Health 1973; 27:168–178.

72. Gamble J, Jones W, Minshall S. Epidemiological-environmental study of diesel bus garage workers: acute effects of NO_2 and respirable particulates on the respiratory system. Environ Res 1987; 42:201–214.

73. Ames RG, Hall DS, Reger RB. Chronic respiratory effects of exposure to diesel emissions in coal mines. Arch Environ Health 1984; 39:389–394.

74. Phelps HW, Koike S. Tokyo-Yokohama asthma. The rapid development of respiratory distress presumably due to air pollution. Am Rev Respir Dis 1962; 86:55–63.

75. Oshima Y, Ishizaka T, Miyamato T, Kabe J, Makino S. A study of Tokyo-Yokohama asthma among the Japanese. Am Rev Respir Dis 1964; 90:632–634.

76. Tremonti LP. Tokyo-Yokohama asthma. Ann Allergy 1970; 28:590–595.

77. Yoshida K, Oshima H, Imai M. Air pollution and asthma in Yokkaichi. Arch Environ Health 1966; 13:763–768.

78. Kitagawa T. Cause analysis of the Yokkaichi asthma episode in Japan. JAPCA 1984; 34:743–746.

79. Imai M, Yoshida K, Kitabatake M. Mortality from asthma and chronic bronchitis associated with changes in sulfur oxide air pollution. Arch Environ Health 1986; 41:29–35.

80. Kagamimori S, Katoh T, Naruse Y, Watanabe M, Kasuya M, Shinkai J, Kawano S. The changing prevalence of respiratory symptoms in atopic children in response to air pollution. Clin Allergy 1986; 16:299–308.

81. Ferris BG, Higgins IT, Peters JM, Ganse WF, Goldman MD. Chronic nonspecific respiratory disease in Berlin, New Hampshire 1961 to 1967. A follow-up study. Am Rev Respir Dis 1973; 107:110–112.

82. Ferris BG, Chen H, Puleo S, Murphy RL. Chronic nonspecific respiratory disease in Berlin, New Hampshire 1967 to 1973. A further follow-up study. Am Rev Respir Dis 1976; 113:475–485.

Chapter 10

Summary and Conclusions

Possible mechanisms

10.1   Many air pollutants found in the UK have, at high concentrations, been shown to compromise the structural or biochemical integrity of the airways and lung and could, in theory, have an effect on asthma. Evidence on mechanisms comes largely from animal and human experimental studies, including human cell culture work. The relevance of this type of evidence to humans exposed to ambient pollution is unclear, and is not sufficient itself to link air pollution with asthma as a public health problem.

[Chapters 4 and 5]

10.2   Experimental work conducted mainly in animals, but also making use of cultured human tissue, has shown that some air pollutants can cause:

- epithelial cell death and denudation of the airway lining;

- increased leakage of fluid from the airway mucosal microvasculature;

- stimulation of inflammatory cell activity;

- release of inflammatory mediators.

[Chapters 4 and 5]

10.3   These pathological effects could affect asthma in the following ways:

- Initiation of asthma:

     direct initiation by exposure to damaging concentrations of toxic substance;

     lowering of the threshold for sensitisation to allergens.

- Provocation of asthma:

     direct provocation by exposure to a toxic substance through an effect on airway responsiveness or lung function;

     lowering of the stimulatory threshold for other provoking factors including allergens and other inhaled irritants.

Human experimental studies

10.4   Evidence from human experimental studies suggests the following:

(1)   Exposure to ozone at ambient concentrations causes airway inflammation and a reduction in lung function. There is some evidence to suggest that ozone may cause a small increase in bronchial responsiveness and may enhance the response to allergen challenge in subjects already sensitised to that allergen.

(2)   Exposure to sulphur dioxide in concentrations sometimes encountered in the UK causes an increase in airway resistance in asthmatic patients. Exposure of non-asthmatic individuals to similar concentrations has little effect.

(3)   Exposure to nitrogen dioxide in concentrations encountered in outdoor air in the UK produces inconsistent effects both in asthmatic and non-asthmatic individuals. The lack of a clear monotonic exposure-response relationship and the wide variation in results obtained in different studies make these data difficult to interpret. There is some evidence to suggest that exposure to nitrogen dioxide may enhance the response to allergen in asthmatic patients, though this effect and its importance remain to be clarified.

(4) Comparatively few laboratory studies of the effects of particles upon the human lung have been undertaken. Studies with acid aerosols at concentrations well above those commonly found in the UK, have shown small responses in normal and asthmatic subjects. There is some evidence to suggest that asthmatic patients are more sensitive to acid aerosols than non-asthmatic subjects.

[Chapter 5]

10.5 Extrapolation to the ambient situation needs to be cautious because of subject selection, qualitative and quantitative differences in exposure to the pollutant and the uncertain significance of outcome measures not associated with symptoms. Most chamber studies involve short exposures to single pollutants often in above-ambient concentrations. This is very different from the ambient situation where exposure to complex pollutant mixtures which fluctuate over time is the rule. In addition, the practice of not involving children, the elderly or severely asthmatic patients in chamber studies means that effects of air pollutants in these groups are poorly understood and possibly underestimated.

Acute effects of air pollution demonstrated in panels of asthmatic and healthy subjects

10.6 Effects are often measured using lung function tests, but the tests usually employed may not reflect acute inflammation of the small airways. Thus, a small effect on lung function may be associated with widespread effects in the small peripheral airways, and absence of an effect on lung function does not exclude toxic damage.

[Chapter 6]

In normal subjects:

10.7 (1) Short-term (eg, daily) variations in ozone are associated with small and temporary reductions in lung function in healthy children and adults. There is individual variability in response and no association with symptoms has been reported. Such reductions in lung function, if superimposed on the already reduced or deteriorating lung function of some asthmatic patients, could provoke symptoms which might not otherwise have occurred.

(2) Current levels of winter pollution (particles associated with SO_2 or NO_2), have less certain effects on healthy persons. Such effects that may occur in healthy persons could cause an increase in symptoms if they were to occur in a person with a pre-existing respiratory condition such as asthma.

[Chapter 6]

Among asthmatic subjects:

10.8 (1) In asthmatic children, most studies indicate that small effects on lung function are likely at the range of concentrations of ozone experienced in the UK, though an increase in symptoms has seldom been reported at these levels.

(2) Similarly, exposure to winter pollution episodes are likely to reduce lung function in asthmatic children by a small amount, but an increase in symptoms is unlikely.

(3) In adult asthmatic patients the effects of exposure to both ozone and winter pollutants are less consistent, but conclusions generally apply as for children.

(4) Though an increase in symptoms has seldom been reported on exposure of asthmatic patients to levels of pollutants as experienced in the UK, it seems likely that patients with very severe asthma will be more affected than mild asthmatic patients.

[Chapter 6]

10.9   The observed effects of air pollution episodes on lung function in both normal and asthmatic subjects are generally small but there appears to be considerable individual variation with some individuals not being affected and others experiencing a change of up to 10% in some indices of lung function. Effects tend to be found at lower levels of the indicator pollutant than would be expected from chamber studies using that pollutant alone. This suggests that other components of pollution have additional effects.

[Chapter 6]

10.10   Most panel studies have involved healthy subjects. Surprisingly few studies have been reported with sufficient numbers of asthmatic patients to allow firm conclusions to be drawn. Extrapolation of the results of panel studies to the general population needs to consider the selection of subjects and patients for the panels and the extent to which the exposure conditions of the study resemble those likely to be encountered in the UK.

Do the long term trends in asthma and other manifestations of allergy correspond to trends in air pollution?

10.11   Trends in asthma and allied allergic disorders are not entirely clear, but available evidence suggests the following:

(1)   In the UK, trends in wheezing illness have been generally upward over the past thirty years. The best data relate to the prevalence among school children and suggest that a relative increase of about 50% may have occurred over that period.

(2)   The cause of this increase is unknown. A number of hypotheses have been proposed, of which air pollution is but one among equally if not more plausible alternatives.

(3)   The relatively greater increases in health services activity for asthma over the past 30 years probably reflect shifts in patterns of diagnosis, clinical management and use of health services.

(4)   The upward trend in hospital admissions shows signs of levelling off.

(5)   There has been a general increase in atopic diseases in industrialised countries during this century.

(6)   It is plausible to link all or part of the increase in asthma prevalence to an increase in atopy (allergic sensitisation). The cause of this is unknown; air pollution is only one of several suggested explanations.

[Chapter 7]

10.12   Trends in both air pollution emissions and measured concentrations in ambient air vary according to the pollutant. There are no satisfactory data available on long-term trends in exposure to ozone or nitrogen dioxide.

(1)   During the past 30 years, there has been a marked fall in emissions of SO_2 and Black Smoke from coal combustion.

(2)   During the past 30 years, there has been an increase in emissions from mobile sources of NO_x and particles (measured as Black Smoke).

(3)   The occurrence of ozone episodes in summer has probably increased over this century, but in the 15–20 years since measurements began, there is no clear trend in annual average concentrations. There is no evidence for an increase in peak (hourly) ozone concentrations over the past 15–20 years. The highest hourly levels observed in the UK in this period were measured in 1976.

(4)   There are only limited data about trends in ambient concentrations of ozone and NO_2 in urban areas.

(5)   In large urban centres such as central London, long term annual average concentrations of NO_2 have not increased, but there are data that suggest that peak levels, measured as a 98th percentile of hourly averages, may be increasing slowly at some urban sites. Diffusion tube surveys in 1986 and 1991 may indicate some increase in ambient concentrations of NO_2 in other urban areas. In contrast, long term annual averages of ozone in urban areas have decreased.

(6)   Measured concentrations of SO_2 continue to fall.

(7)   Measured concentrations of Black Smoke have been fairly stable since 1980 with some evidence of a slow downward trend. The decline in emissions of particles from fixed sources is countered by the increase in emissions from vehicles.

(8)   Black Smoke levels remain higher in urban than in rural areas.

(9)   For some pollutants (eg, NO_2) indoor air makes an important contribution to personal exposure. Little is known about trends, but it has been suggested that indoor pollution could have increased as a result of better draught proofing of houses.

(10) The dose of air pollutants to the lung will also depend on factors such as the amount of time spent in various environments and the level of activity. Nothing is known about trends in these factors.

[Chapter 7]

10.13   There is no consistent correlation between time trends in air pollution and those in asthma. There is a negative relationship with pollution from coal combustion (Black Smoke and SO_2). This argues against Black Smoke and SO_2 playing a role in the initiation of asthma. In contrast, there is a positive correlation with the increase in vehicle-related emissions, though not necessarily with exposure. In considering these trends it should be noted that the increase in some indicators of asthma may have slowed or stopped.

10.14   There are few studies which have convincingly demonstrated an increase in asthma in association with new sources of pollution. Similarly, there is no convincing evidence of a decline in asthma prevalence in areas which have experienced an improvement in ambient air quality.

[Chapter 7]

10.15   There are, however, other factors which could be linked to the increase in prevalence of asthma, including exposure to indoor chemical air pollutants and allergens, maternal smoking, exposure to infection and composition of the diet. These have also changed, or are likely to have changed, in the United Kingdom over recent decades. Thus, there are other plausible explanations for trends in atopy and asthma.

Do seasonal patterns of asthma correspond to seasonal variations in pollution?

10.16   Seasonal patterns of asthma are complex and bear little relation to cyclical variations of major non-biological air pollutants. Alternative and more plausible explanations exist for seasonal variations in asthma including the seasonal variations in exposure to pollens, viral infections and weather conditions.

[Chapter 8]

Are daily variations in asthma morbidity associated with daily variations in air pollution?

10.17   The question of daily variations in asthma morbidity has been addressed mainly by time-series studies which use grouped data available for populations through routine systems for recording service utilisation. Analysis is usually of daily events, but deaths from asthma are too infrequent for this. Studies of sulphur dioxide, the pollutant to which asthmatic patients are most clearly sensitive in chamber studies, show no evidence of an association with daily asthma admissions. There is suggestive evidence of an adverse effect of winter particulate pollution on asthma admissions in the US, but UK data are too limited to assess whether this applies in the UK. Nitrogen dioxide levels are unlikely to be an important influence on daily variations in asthma occurrence.

[Chapter 8]

10.18 Evidence relating asthma to summer pollutants, such as ozone, is complex and contradictory. Several studies of populations in eastern USA and Ontario have found higher asthma admission rates on high ozone days, whereas no such correlation, or an inverse relationship, has been reported from the west coast of America. The summertime pollutant mixtures in west and east America may be significantly different from each other and from the mixture occurring in the UK and the reasons for the variations in the results have yet to be established. For these reasons, and because there is no published time-series analysis relating to ozone in the UK, there is insufficient evidence to draw a firm conclusion regarding possible effects of ambient ozone levels on the timing of asthma attacks in Britain.

[Chapter 8]

Are asthma epidemics caused by air pollution?

10.19 Asthma epidemics undoubtedly occur, but where a cause has been identified, this has always been an increase in aeroallergen levels, with little or no proven contribution from non-biological air pollution. Asthma epidemics have also been associated with thunderstorms in the UK, but are unlikely to have been due to non-biological air pollutants.

[Chapter 8]

Are air pollution episodes associated with an increase in asthma?

10.20 With few exceptions, episodes of urban "smog" (Black Smoke and SO_2) in the past have not been notable for their effect on asthma in children or young adults. The main effects have been on older subjects with chronic respiratory and cardiovascular disease.

[Chapter 8]

10.21 In the 1991 smog episode in London associated with high NO_2 and Black Smoke levels, asthma admissions and consultations were not significantly increased compared with other respiratory conditions, nor relative to asthma admissions and consultations in the surrounding south east of England.

[Chapter 8]

Do regional variations in air pollution correspond to regional variations in asthma?

10.22 The regional pattern of asthma in the UK is poorly defined. Within England and Wales, the patterns of asthma prevalence, utilisation of health services and mortality bear little relationship to regional variations in the main pollutants of current concern.

[Chapter 7]

Does the geographical distribution of asthma prevalence show a positive association with levels of air pollution?

10.23 The geographical distribution of asthma has been studied in relation to levels of outdoor air pollutants, but most of the available studies are deficient in statistical power, having too few (usually two) units of comparison. Most studies are relevant to smoke and SO_2 pollution. The results are inconsistent and most of those from Western Europe and America have not found substantial relationships between pollution and asthma. Those within Eastern Europe generally find an association with life-time prevalence of wheezing in children. There is less evidence about the geographical association of asthma with the newer forms of pollution, but those studies which are available do not provide compelling evidence that the two are related. Two cohort studies from California raise the possibility that the incidence of asthma may be related to levels of long term exposure to photochemical oxidants.

[Chapter 9]

10.24   Prevalence studies clearly indicate that indoor sources of pollution, especially tobacco smoke, are significant factors in wheezing illness, particularly in children. It is not clear whether indoor exposure affects mainly the initiation or provocation of asthma. Where indoor air pollution has been taken into account, its influence has usually been greater than outdoor air pollution.

[Chapter 9]

Is the prevalence of asthma higher in urban than in rural environments?

10.25   There is little evidence of variation of asthma prevalence or mortality by degrees of urbanisation whether in urban populations exposed to old or new patterns of pollution.

[Chapter 9]

10.26   Migrants from underdeveloped to developed environments have been found to have increased asthma or bronchial hyperresponsiveness. This could be explained by exposure to elements of western lifestyle, urban air pollution being only one of several possible factors.

[Chapter 9]

Is the prevalence of asthma related to exposure to traffic?

10.27   In most of the few studies so far published, there is a consistent, though modest, association between exposure to traffic and asthma prevalence in children. Assuming that this association is real (and not due to reporting bias, for example) it is unclear whether this is due to an increase in the initiation or provocation of asthma. Air pollution is a plausible explanation though socio-economic and other environmental factors could also play important roles.

[Chapter 9]

Are variations in prevalence within the united Germany consistent with an adverse effect of the "new" pollution on asthma and allergic disease?

10.28   Studies comparing former East and West Germany provide an opportunity to examine cross-sectionally the effects of old and new patterns of pollution within genetically similar populations. Most studies were, however, done after there had already been a major fall in pollution in Eastern Europe.

[Chapter 9]

10.29   Generally, lower levels of skin test allergy have been found in the former East Germany than in the former West Germany. Total IgE concentrations in blood are, in contrast, higher in the East. Bronchitis seems more common and asthma less common in the East. There is no difference in the levels of bronchial hyperresponsiveness. The role of air pollution in explaining these phenomena is uncertain.

[Chapter 9]

Conclusions

10.30   As regards initiation of asthma most of the available evidence does not support an effect of non-biological air pollution.

10.31   As regards worsening of symptoms or provocation of asthmatic attacks, most asthmatic patients are unaffected by exposure to such levels of non-biological air pollution as commonly occurs in the UK. A small number of patients experience clinically significant effects which occasionally require an increase in medication or attention by a doctor.

10.32   Factors other than air pollution are influential with regard to the initiation and provocation of asthma and are much more important than air pollution in both respects.

10.33   Asthma has increased in the UK over the past thirty years but this is unlikely to be the result of changes in air pollution.

Appendix 1

Recommendations for Further Research

A1.1   Research recommendations include: epidemiological studies, chamber or volunteer studies and work involving *in vivo* and *in vitro* techniques.

Epidemiological studies

A1.2   Trends in asthma and atopic sensitivity should be studied. This could be done in two ways: 1) retrospectively, by repeating past studies using identical methods, or 2) prospectively, to monitor future trends. Emphasis should be on studies of large samples, or with nationwide coverage.

A1.3   Changes in the prevalence of asthma should be compared with changing patterns of air pollution. Such studies would need to consider changes in other environmental factors which could also affect the incidence or prevalence of asthma.

A1.4   Possible relationships between the incidence of asthma and atopic sensitivity and air pollution should be investigated by means of cohort studies. These could be retrospective, using existing cohorts and would require estimation of life-time exposure to pollutants. Prospective cohort studies should also be considered, including cohorts starting in the antenatal period.

A1.5   Time-series analyses of emergency health service contacts for asthma in relation to daily fluctuations in levels of air pollutants, within the UK, with rigorous adjustment for seasonal and meteorological factors and, where possible, aeroallergen levels, are recommended.

A1.6   Panel studies of daily symptoms, lung function and medication requirements among samples of asthmatic patients in the UK, representing a range of ages, disease severity and atopic status. These studies could usefully incorporate comparisons of daily variations in personal exposure to gaseous and particulate pollution.

A1.7   Multi-centre cross-sectional studies comparing asthma prevalence and severity and atopic sensitivity in a large number (>20) of areas of well characterised but differing mean air pollutant concentrations, in the UK or elsewhere, with information collected about the major sources of indoor air pollution.

A1.8   Analyses of asthma prevalence, severity and health service utilisation in urban areas of the UK in relation to local traffic density, including estimates of pollutant exposure and adjustment for socioeconomic factors and indoor pollution are recommended.

Chamber or volunteer studies

A1.9   Volunteer studies designed to identify the nature of the inflammatory response produced in patients with asthma on exposure to air pollutants. Subjects in such studies should be as representative as possible of asthma patients in general.

A1.10   Studies of early adaptive responses to air pollutants and of whether the extent of adaptation varies significantly between those suffering from asthma and other individuals are recommended.

A1.11   Further studies on the effects on asthmatic patients of exposure to low concentrations of nitrogen dioxide.

A1.12   Studies of the effects of exposure to air pollutant on the response of sensitised individuals to allergens should be continued.

In vivo and in vitro laboratory studies

A1.13   The need for a satisfactory animal model for asthma remains. Without such a model, mechanistic studies are made more difficult. Further work on the development of such a model is recommended.

A1.14   The effects of air pollutants on the function of basal cells of the proximal and distal airways as anchor cells of the epithelium require further study.

A1.15   Studies of the adjuvant effects of air pollutants (especially particles) on the response to inhaled allergens are recommended.

Appendix 2

Glossary of Terms and Abbreviations

Acetylcholine	Neurotransmitter between nerve cells and at the neuromuscular junction
Aeroallergens	Airborne antigens that cause allergy in a sensitive individual, eg, pollen, house dust mite
AHR	Airway hyperresponsiveness
Aldehydes	A group of chemical compounds several of which are irritant, eg, formaldehyde
Allergic rhinitis	"Hay fever", "perennial rhinitis"
Allergic sensitisation	The result of first exposures to an antigen which induce an immune response, so that on later exposure to the same antigen an allergic response will occur
Alternaria tenuis	A mould (fungus)
Alveolus	Smallest division of the lung, a thin-walled sac surrounded by blood vessels, through which the respiratory exchange of gases occurs
Anaphylactic reaction	Hypersensitivity reaction, leading to anaphylactic shock, which may be fatal
Anergy	Diminished hypersensitivity to specific antigen(s)
Antigen-presenting cells	Cells which present processed antigen molecules to T-lymphocytes. Includes macrophages, M cells of gut and dendritic cells
API	Airborne particulate index
Ascaris suum	A species of *Ascaris*, a parasitic worm found in swine
Atopy	A tendency to develop unduly large amounts of allergic antibodies (IgE) to non-infectious environmental agents
Autoregressive modelling	Regression modelling taking into account that measurements on a given day may depend on values on preceding days
BAL	Bronchoalveolar lavage
Basal cells	Epithelial (qv) cells next to the basement membrane (qv) believed to be important in anchoring other epithelial cells
Basement membrane	A thin membrane that supports an overlying epithelium or endothelium
Beta blockers	Drugs which block the action of transmitters at β receptors: used to control blood pressure and reduce heart rate
bFGF	Fibroblast Growth Factor: a cytokine
BHR	Bronchial hyperreactivity: the tendency to develop bronchospasm on exposure to a range of "non-specific" stimuli, including exercise, cold air, histamine or methacholine
Black Smoke	Non-reflective (dark) particulate matter, associated with the smoke stain measurement method. Generally less than 4.5 μm diameter
Bordetella pertussis	The bacterium that causes whooping cough
Bronchial mucosa	The mucous membrane of the airways
Bronchiectasis	A disease characterised by dilation of the bronchi and purulent sputum

Bronchiole	A small air passage (generally <1 mm in diameter) and lacking cartilage and glands
Bronchiolitis	Inflammation of the bronchioles
Bronchitis	Inflammation of the bronchi. Also the clinical condition caused by such inflammation
Bronchospasm	Narrowing of bronchi by muscular contraction
Bronchus	(Plural: bronchi). Major air passages beyond the trachea, with cartilage and mucous glands in walls
Brownian motion	Constant random movement of suspended bodies due to bombardment by surrounding molecules
Capsaicin	An extract from chilli peppers which causes bronchospasm (qv)
Causality	A direct link between an exposure and an effect
CD4$^+$, CD8$^+$, CD25$^+$	Cell surface proteins on T-lymphocytes which act as specific receptors (CD: cluster differentiation)
C-fibres	Fine naked nerve endings found close to the surface of epithelium which respond to noxious stimuli
CGRP	Calcitonin gene-related peptide: a cytokine
Chamber studies	Studies involving exposure of volunteers to controlled atmospheres in experimental chambers
Chemotactic factors	Factors which cause movement of a cell along a gradient of chemical concentration
CHESS	Community Health and Environmental Surveillance System (USA)
Cilia	Fine, hair-like projections on the surface of respiratory epithelial cells which beat in unison to clear mucus and inhaled particles
Ciliary dyskinesia	Congenital defect of cilia
CI	Confidence interval
Clara cells	Unciliated cells found mainly in the smaller airways: important in metabolising inhaled toxins
Clones	A group of genetically identical cells derived from a single progenitor
CMD	Count Median Diameter. Term used to characterise the size distribution of particles in an aerosol. 50% of particles are of diameter less than the count median diameter
Coefficient of haze	A measure of the optical absorption density of airborne particles, closely related to Black Smoke
Cohort	A defined group of persons who are followed or traced over a period of time
COMEAP	Committee on the Medical Effects of Air Pollutants
Complement	Group of proteins involved in inflammation and cell destruction
COPD	Chronic obstructive pulmonary disease
Cytokines	Proteins which stimulate or inhibit the differentiation, proliferation or function of immune cells and other cells. Have autocrine/paracrine activity
Dendritic antigen presenting cells	Cells with long finger-like processes, which present processed antigen and MHC (qv) to T-cells
DEP	Diesel exhaust particles
Dermatophagoides pteronyssinus	The commonest house dust mite

Desmosomes	Intercellular junctions characterised by specific binding proteins
DTPA	^{99m}Tc-diethylene triaminopentaacetate
EAR	Early airway response
EEC	European Economic Community
Eicosanoid	Generic term embracing prostaglandins, thromboxanes and leukotrienes derived from arachidonic acid, a 20 carbon molecule
ELF	Epithelial lining fluid
Endothelins	Factors released which play a role in controlling vascular smooth muscle. First demonstrated in endothelial cells
Endothelium	Single layer of cells lining blood vessels or lymph vessels
Eosinophils	White blood cells containing distinctive granules involved in the immune response
Eosinophil cationic protein	A protein present in the eosinophil granule matrix
Eosinophil-derived neurotoxin	Neurotoxic material found in eosinophils
Eosinophil peroxidase	Enzyme present in the eosinophil granule matrix which catalyses oxidation
EPA	Environmental Protection Agency (US)
Epidemiology	The study of the distribution and causes of disease in populations
Epithelial cell-derived relaxing factors	Factors released from epithelial cells in response to activation of certain receptors, producing relaxation of smooth muscle
Epithelium	Layer of cells covering the surfaces of the body (the skin and mucous membranes and the cavities of the body). In the latter the layer is referred to as a mesothelium
E-selectin	An adhesion molecule on endothelial cells which facilitates the adhesion of other cells, e.g., of leukocytes to the endothelium
extra-Poisson variation	For data expressed as counts, dispersion where the variance of the data is significantly greater than the average (mean) value
FEF$_{25-75}$	The mean forced expiratory flow rate during the middle half of a forced expiration from full inspiration (l.sec^{-1})
FEV$_1$	The volume of air expired during the first second of a maximal or "forced" expiration
Fibronectin	An extracellular protein that supports and connects tissue cells. Also promotes differentiation and cytoskeletal organisation
Fume	Fine particles suspended in air. Particle diameters 1 nm to 1 μm
FVC	Forced vital capacity. The volume of air expired in a forced expiration following maximum inspiration
Gaussian distribution	A smooth, symmetrical bell-shaped probability curve, otherwise known as the normal distribution
Gaussian errors	The representation of the random or unexplained part of a regression model as following the Gaussian (Normal) distribution
GDR	Former German Democratic Republic
GM-CSF	Granulocyte-macrophage-colony-stimulating factor: a cytokine
Gravimetric analysis	Analysis of materials and estimation of constituents by weight
H$^+$	Hydrogen ion
HCl	Hydrochloric acid
Heparan sulphate	Proteoglycan component of connective tissue ground substance

HES	Hospital Episode System
15-HETE	15-Hydroxyeicosatetraenoic acid, a pro-inflammatory metabolite of arachidonic acid
HIPE	Hospital Inpatient Enquiry
Histamine	Compound involved in inflammation and allergic reactions, released by mast cells
Histopathological	Pertaining to changes in tissue structure due to disease or insult
H_2O_2	Hydrogen peroxide
H_2SO_4	Sulphuric acid
5HT	5-hydroxytryptamine or serotonin: a vasoconstrictor substance, also a neurotransmitter
Hyperplasia	Increase in the number of cells, causing an increase in the size of an organ or tissue
$h\upsilon$	Energy of electromagnetic radiation
HLA-DR	One of the genetic loci involved in typing the human leukocyte antigen, the human MHC (qv)
ICAM–1	Intercellular adhesion molecule 1
IFN–γ	An interferon involved in the regulation of the immune response, a cytokine
IgA	Immunoglobulin A, an antibody
IgE	Immunoglobulin E, the allergy antibody
IgG	Immunoglobulin G, an antibody
IL–2 to IL–13	Interleukin-2 to Interleukin 13, cytokines (qv) secreted by leukocytes
Immunopathology	The study of abnormal immune responses
Incidence	The frequency with which cases of a given disease present in a particular period for a given population. Incidence (persons) refers to the rate of development of new cases. Incidence (spells) refers to the rate of occurrence of episodes of ill-health
Indicator pollutants	Pollutant chosen to characterise an air pollution episode, eg, ozone in summer episodes of photochemical air pollution
Indomethacin	A non-steroidal anti-inflammatory drug
Initiators (of asthma)	Factors which lead to the establishment of the asthmatic state
Integrin $\alpha_6\beta_4$	One of a group of molecules concerned in T-lymphocyte functions, a cell surface adhesion molecule
LAR	Late airway response
Leukotrienes (LTs)	Molecules which promote inflammatory processes, metabolites of arachidonic acid
Ligand	In a complex chemical structure the ions, atoms or molecules, surrounding the central atom. Used to refer to molecules binding to receptors
LRT	Lower respiratory tract
LTB_4	Leukotriene B_4
LTC_4	Leukotriene C_4
LTD_4	Leukotriene D_4
LTE_4	Leukotriene E_4
Lung function tests	Tests of the ventilatory or mechanical functions of the lung
Lymphocytes	White blood cells which produce antibodies or chemical messengers concerned in immune reactions

Lymphokines	Cytokines (qv) produced by lymphocytes
Lysosomal enzymes	Enzymes involved in the digestion of material phagocytosed by cells
Macrophages	Phagocytic scavenger cells of the tissues as opposed to those of the blood, secrete cytokines and proteases
Mast cells	Cells which when cross-linked by antigen to IgE mediate an immune response. These cells produce histamine and cytokines
MHC	Major histocompatibility complex. A genetic region encoding molecules involved in antigen presentation to T-lymphocytes
MBP (Major basic protein)	A protein stored in eosinophils, which is highly toxic to bronchial and alveolar epithelial cells and also induces hyperresponsiveness at high concentrations
MCh	Methacholine
Methylprednisolone	A synthetic corticosteroid used to inhibit inflammation and hypersensitivity
Micrometre (μm)	A thousandth of a millimetre
MMD	Mass Median Diameter
MMEF	Maximal mid-expiratory flow
Monotonic	A trend of varying size, but always in one direction
Mucin	A viscous glycoprotein occurring in the mucus of saliva and other secretions
Myofibroblasts	Contractile connective tissue cells
Nedocromil sodium	Drug with a pharmacological action similar to that of sodium cromoglycate used for prophylaxis of asthma
Neutral endopeptidase (NEP)	An enzyme that modulates neuropeptidase activity by proteolytic activation
Neurokinin A	A tachykine neuropeptide
Neutropenic	Abnormal decrease in the number of neutrophil leukocytes in the blood
NH_3	Ammonia
NILU	Norwegian Institute for Air Research
NFκB	A transcription factor that regulates expression of genes involved in the inflammatory response eg cell adhesion molecules and cytokines
nmol	Nanomole, 10^{-9} mole. [A mole is the relative molecular mass in grams]
NO	Nitric oxide, properly called nitrogen monoxide
NO_2	Nitrogen dioxide
NO_x	Total oxides of nitrogen. Conventionally the mixture of NO and NO_2 in the atmosphere
O_3	Ozone
Ontogeny	Development of the individual or of specified cells
Opsonisation	Enhancing phagocytosis by promoting adhesion of an antigen to a phagocyte
Ovalbumin	Soluble protein from egg white
PAARC	Pollution Atmospherique et Affections Respiratoires Chroniques (France)
PAF	Platelet activating factor
Panel studies	Epidemiological method in which a cohort of individuals is followed prospectively over time to examine the associations between air pollution and health indicators, usually measured on a daily basis

PC$_{20}$	Provocation concentration 20: the dose of a test compound which causes a 20% fall in FEV$_1$
PEFR	Peak expiratory flow rate
Platelet derived growth factor (PGDF)	A cytokine originally isolated from platelets but now known to be located in other cells
PGD$_2$, PGE$_2$, PGF$_2$	Prostaglandin D$_2$, prostaglandin E$_2$, prostaglandin F$_2$ (metabolites of arachidonic acid)
Phenotype	The observable physical, biochemical and physiological characteristics of a cell, tissue, organ or individual
Photochemical smog	Smog caused by the formation of particles and gases due to a chemical reaction driven by sunlight
PM$_{2.5}$	Particulate matter less than 2.5 μm aerodynamic diameter (or, more strictly, particles which pass through a size selective inlet with a 50% efficiency cut-off at 2.5 μm aerodynamic diameter)
PM$_{10}$	Particulate matter less than 10 μm aerodynamic diameter (or, more strictly, particles which pass through a size selective inlet with a 50% efficiency cut-off at 10 μm aerodynamic diameter)
Poisson errors	For data expressed as counts, variation described as a Poisson distribution, implying that variance is the same as the mean value
ppb	Parts per billion, 1 part by volume in 10^9
ppm	Parts per million, 1 part by volume in 10^6
Prevalence	The proportion of persons affected at a particular time (point prevalence) or over a stated period (period prevalence)
Prostanoids	Analogues of prostanoic acid (prostaglandin)
Pseudostratified epithelium	An epithelium appearing to comprise layers of cells in which all cells are actually in contact with the basement membrane
Pulmonary oedema	Accumulation of fluid within the air spaces of the lung
RANTES	Regulated upon Activation, normal T-cell expressed and secreted: a cytokine
Raw	Airway resistance
Regression model	A basic methodology of modern applied statistics expressing the expected value of outcome data as a function of explanatory variables; and the random or unexplained variation of individual outcome data points about that average as a defined statistical distribution
Ribonuclease	Enzyme which cuts ribonucleic acid
RR	Relative risk: the ratio of the incidence or prevalence in two groups
Secretory leukoprotease inhibitor	An inhibitor of neutrophil elastase and other serine proteases; synthesised and secreted by non-ciliated bronchiolar epithelial cells
Sensory neuropeptides	Peptides, including substance P, neurokinins A and B and calcitonin gene related peptide found in mucosal sensory nerves. These are released on stimulation of nerve endings and produce a local inflammatory response
Smog	A term used to describe a mixture of smoke and fog. Also used to describe photochemical air pollution
SOB	Shortness of breath
SO$_2$	Sulphur dioxide
SO$_4$	Sulphate

Squamous metaplastic cells	Abnormal (flat) cells produced in the airways as a result of continued irritation
Status asthmaticus	An attack of asthma with an increased risk of death from respiratory failure or exhaustion
Substance P	A sensory neuropeptide which causes most smooth muscle to contract, but vascular smooth muscle to relax
Summer haze pollution	See photochemical smog
Suppressor T-lymphocytes	Subpopulation of lymphocytes originating in the thymus (hence T-lymphocytes) which are capable of suppressing antibody production or a specific cell-mediated response to an antigen
Surfactant	Agent which reduces surface tension. Lung surfactant (produced by Type II alveolar cells and Clara cells) contributes to the elasticity of the tissue
Tachykinins	A group of peptides, of which the major example is substance P (qv)
T Helper cells	Subpopulation of lymphocytes originating in the thymus (hence T-lymphocytes) which recognise and bind to antigen plus MHC (qv) molecules and produce cytokines (qv) which switch on the immune response
Th1, Th2	Two subpopulations of T helper cells (qv)
TNF–α, TNF–β	Tumour necrosis factor–α, tumour necrosis factor–β, two related cytokines (qv), originally named for their cytotoxic effects on tumour cells, but which also have immunoregulatory functions
TNT	Trinitrotoluene
Tryptase	Proteolytic enzyme synthesised and secreted by mast cells
Total suspended particulate (TSP)	A term describing the gravimetrically determined mass loading of airborne particles, most commonly associated with the use of the US high volume air sampler in which particles are collected on a filter for weighing
URT	Upper respiratory tract
US	United States
Uvomorulin	A cell adhesion molecule that straddles the cell membrane to form tight junctions and confers adhesive specificity during tissue development and repair
VCAM-l	Vascular cell adhesion molecule
VIP	Vasoactive intestinal peptide, a sensory neuropeptide
VOCs	Volatile organic compounds
WHO	World Health Organization
ZnO	Zinc oxide

Concentration Units and Conversion Factors

Concentrations of air pollutants are expressed in two ways, either as the mass of pollutant in a given volume of air (usually expressed as micrograms per cubic metre or $\mu g/m^3$) or as the ratio of the volume of the gaseous pollutant (expressed as if pure) to the volume of air in which the pollutant is contained (usually expressed as a volume mixing ratio or parts per million, ppm, or parts per billion, ppb).

The mass concentration as expressed above will be dependent on the ambient temperature and pressure and ideally these should be specified each time a concentration is measured as a mass/volume. The variation is discussed below and although not large may not be negligible where large variations in temperature and pressure occur.

The volume mixing ratio is independent of temperature and pressure, if ideal gas behaviour is assumed.

The relationship between the two sets of units can be expressed as follows:

$$\mu g/m^3 = ppb \times \frac{molecular\ weight}{molecular\ volume}$$

where:

$$molecular\ volume = 22.41 \times \frac{T}{273} \times \frac{1013}{P}$$

where T is the ambient temperature (°K) and P is the atmospheric pressure (in millibars). Conversion factors for some common gaseous pollutants are given in the Table below for 20°C and 0°C and 1013 mb pressure. Pollutants which are present in particulate form in the atmosphere such as sulphates are normally only expressed in mass/volume units.

| Pollutant | Molecular weight | To convert | | | |
| | | ppb to $\mu g/m^3$ | | $\mu g/m^3$ to ppb | |
		0°C	20°C	0°C	20°C
NO_2	46	2.05	1.91	0.49	0.52
NO	30	1.34	1.25	0.75	0.80
HNO_3	63	2.81	2.62	0.36	0.38
O_3	48	2.14	2.00	0.47	0.50
SO_2	64	2.86	2.66	0.35	0.38
CO†	28	1.25	1.16	0.80	0.86

* ie, to convert ppb of SO_2 at 0°C to $\mu g/m^3$ multiply by 2.86

†for CO the factors apply to the more commonly used conversions of ppm and mg/m^3

Appendix 3

Membership of Committee on the Medical Effects of Air Pollutants

Chairman	Professor S T Holgate, MD, DSc, FRCP, FRCPE
Members	Professor H R Anderson, MD, MSc, FFPHM
	Dr J G Ayres, BSc, MD, FRCP
	Professor P G J Burney, MA, MD, MRCP, FFPHM
	Dr M L Burr, MD, FFPHM
	Professor R L Carter, MA, DM, DSc, FRCPath, FFPM
	Professor B Corrin, MD, FRCPath
	Professor A Dayan, MD, FRCP, FRCPath, FFPM, CBiol, FIBiol
	Professor R K Griffiths, BSc, MB, ChB, FFCM
	Professor R M Harrison, PhD, DSc, CChem, FRSC, FRMetS, FRSH
	Dr D Purser, BSc, PhD
	Dr R J Richards, BSc, PhD, DSc
	Professor A Seaton, MD, FRCP, FFOM
	Professor A E Tattersfield, MD, FRCP
	Mr R E Waller, BSc
	Dr S Walters, BSc, MRCP, MFPHM
Secretariat	Dr R L Maynard, BSc, MB, BCh, MRCPath, CBiol, FIBiol *(Medical)*
	Mrs A McDonald, BSc, MSc *(Scientific)*
	Miss J P Cumberlidge, BSc, MSc *(Minutes)*
	Mr J D Raghunath *(Administrative)*

Appendix 4

Membership of Sub-Group on Asthma and Outdoor Air Pollution

Chairman	Professor H R Anderson, MD, MSc, FFPHM
Members	Dr J G Ayres, BSc, MD, FRCP
	Professor A J Newman Taylor, OBE, MSc, FRCP, FFOM
	Dr D P Strachan, MSc, MD, MRCP, MFPHM, MRCGP
	Dr T D Tetley, BSc, PhD
	Professor J O Warner, MD, DCH, FRCP

Printed in the United Kingdom for HMSO
Dd 0301582   10/95   65536   3400   335885   40/33740

Abnormal Psychology